Springer Theses

Recognizing Outstanding Ph.D. Research

Aims and Scope

The series "Springer Theses" brings together a selection of the very best Ph.D. theses from around the world and across the physical sciences. Nominated and endorsed by two recognized specialists, each published volume has been selected for its scientific excellence and the high impact of its contents for the pertinent field of research. For greater accessibility to non-specialists, the published versions include an extended introduction, as well as a foreword by the student's supervisor explaining the special relevance of the work for the field. As a whole, the series will provide a valuable resource both for newcomers to the research fields described, and for other scientists seeking detailed background information on special questions. Finally, it provides an accredited documentation of the valuable contributions made by today's younger generation of scientists.

Theses are accepted into the series by invited nomination only and must fulfill all of the following criteria

- They must be written in good English.
- The topic should fall within the confines of Chemistry, Physics, Earth Sciences, Engineering and related interdisciplinary fields such as Materials, Nanoscience, Chemical Engineering, Complex Systems and Biophysics.
- The work reported in the thesis must represent a significant scientific advance.
- If the thesis includes previously published material, permission to reproduce this must be gained from the respective copyright holder.
- They must have been examined and passed during the 12 months prior to nomination.
- Each thesis should include a foreword by the supervisor outlining the significance of its content.
- The theses should have a clearly defined structure including an introduction accessible to scientists not expert in that particular field.

More information about this series at http://www.springer.com/series/8790

Xu Wang

Energy Consumption, Chemical Use and Carbon Footprints of Wastewater Treatment Alternatives

Assessment Methodology and Sustainability Solutions

Doctoral Thesis accepted by
the Research Center for Eco-Environmental Sciences, Chinese Academy of Sciences, Beijing, China

Springer

Author
Dr. Xu Wang
Research Center for Eco-Environmental Sciences
Chinese Academy of Sciences
Beijing, China

Supervisor
Prof. Junxin Liu
Research Center for Eco-Environmental Sciences
Chinese Academy of Sciences
Beijing, China

ISSN 2190-5053 ISSN 2190-5061 (electronic)
Springer Theses
ISBN 978-981-13-5982-8 ISBN 978-981-13-5983-5 (eBook)
https://doi.org/10.1007/978-981-13-5983-5

Library of Congress Control Number: 2019934799

This Springer imprint is published by the registered company Springer Nature Singapore Pte Ltd.
The registered company address is: 152 Beach Road, #21-01/04 Gateway East, Singapore 189721, Singapore

Supervisor's Foreword

It is clear that wastewater services and water management, in general, must transition to become much more resource-efficient to continue to serve the needs of society in the future. Because many wastewater service and water management infrastructure elements have service lives of 50 to 100 years or more, a decision made today has long-lasting implications and must be made based on future rather than current or past scenarios.

Fortunately, leading experts are recognizing this, and many new concepts and technologies are becoming available, which can be truly transformative. However, the evaluation of these new concepts and technologies is difficult because they are components of larger systems and are intended to achieve a broader range of societal objectives. Public health and environmental protection objectives have been sufficient justifications and bases for the evaluation of alternative approaches in the past, but future systems must provide a broader range of environmental benefits, along with social and economic benefits. Consequently, new methodologies to evaluate options based on this expanded range of criteria that truly focus on the selection of the most sustainable option are needed.

Accordingly, Dr. Wang's research was characterized by a unique combination of original thinking about systems (in this case, wastewater treatment systems) and analytical skills, which helped him to compare and evaluate options in an integrated and comprehensive fashion. In his thesis, Dr. Wang envisioned and conceptualized the entire system while focusing, as necessary, on its specifics. He used appropriate and highly innovative analysis tools to perform comprehensive assessments, which will be beneficial for envisioning and developing new systems that are needed as we transition from our historical resource consumptive economic base to a circular economy with future sustainability.

The transformations required for water and wastewater management infrastructure are massive, global in scope, and essential to human well-being and survival. They require our best thought and analysis to identify and plot the right course forward. The core of this thesis is a key component to solving this challenge.

Beijing, China
October 2018

Prof. Junxin Liu

Abstract

Wastewater treatment systems are receiving attention because of their higher electricity consumption, increased excess sludge production, and elevated greenhouse gas (GHG) emissions. However, current estimation schemes for wastewater treatment systems primarily focus on the treatment efficiency, effluent quality, and environmental consequences for receiving water bodies. These factors can generally not be used to quantify the potential conversion of wastewater pollutants into recoverable resources. Therefore, a refined evaluation scheme is proposed in this study, which quantifies adverse environmental effects in addition to bioenergy and nutrient recovery indices. A novel regulation pattern for pollutant transformation is proposed to facilitate energy conservation, emission reduction, and resource recovery during the wastewater treatment process.

The initial investigation was conducted to construct and evaluate six anaerobic/anoxic/oxic wastewater treatment systems, which were designed to meet various treatment standards in China, from a life cycle perspective. The results reveal that the sophisticated treatment proposed for local receiving waters can be realized at the expense of higher energy consumption, chemical use, and GHG emissions. However, the lower environmental sustainability of the wastewater treatment mentioned above can be mitigated by resource recovery options, even if increasingly stringent discharge standards are required.

Several estimation indicators are included in the proposed evaluation scheme. Each indicator has a specific significance, which may substantially affect the final evaluation results. Therefore, the significance of each estimation indicator was weighted carefully using a global time series database. Data for 52 countries collected between 1990 and 2009 were summarized and analyzed. The outcomes reveal that the weighted coefficients of each indicator are relative and vary by period and country. The weighted indicator data highlight the factors that are important for a certain period. Generally, the differences in the weighted indicator values between developing and developed countries are remarkable and exhibit regularity. For example, the tendency of the weighted data of the bioenergy recovery indicator in developing countries between 2000 and 2009 is similar to that in developed nations between 1990 and 2000. This reveals that the weight coefficients of several indicators in developing countries are merely behind those of developed countries. Forecast modeling for the weighting of each indicator is proposed based on statistical principles. Based on scenario analysis and prediction, phosphorus capture via struvite generation will be the most sustainable option for developed countries in the future. Nutrient recovery via sludge composting will be the better choice in developing countries.

The results also indicate that carbon substrates will significantly affect the energy consumption efficiency and resource recovery potentials of the wastewater treatment process. Furthermore, acetic acid is dominated by substrate storage in a traditional activated sludge process, rather than by total oxidization. This results in an energy reduction in the water processing line and facilitates phosphorus gain.

Finally, based on the present regulation pattern for pollutant transformation, the effluent has the potential to meet Chinese Class 1A standards. Compared with the traditional wastewater treatment pattern, 62.3% and 41.7% of the energy consumption and GHG emission, respectively, can be reduced with the proposed regulation pattern. Based on the proposed pattern, the treatment of one cubic meter of wastewater produces a net energy gain of 0.34 kWh.

Keywords Wastewater treatment · Energy consumption · Chemical use · Greenhouse gas emission · Mitigation · Resource recovery · Integrated assessment model · Scenario analysis · Regulation pattern · Environmental sustainability

Parts of this thesis have been published in the following journal articles

Xu Wang, Glen Daigger, Duu-Jong Lee, Junxin Liu, Nan-Qi Ren, Jiuhui Qu, Gang Liu, David Butler. Evolving wastewater infrastructure paradigm to enhance harmony with nature. *Science Advances* 2018, 4, eaaq0210. (Reproduced with Permission).

Xu Wang, Perry L. McCarty, Junxin Liu, Nan-Qi Ren, Duu-Jong Lee, Han-Qing Yu, Yi Qian, Jiuhui Qu. Probabilistic evaluation of integrating resource recovery into wastewater treatment to improve environmental sustainability. *Proceedings of the National Academy of Sciences of the United States of America* 2015, 112(5), 1630–1635. (Reproduced with Permission).

Xu Wang, Junxin Liu, Bo Qu, Nan-Qi Ren, Jiuhui Qu. Role of carbon substrates in facilitating energy reduction and resource recovery in a traditional activated sludge process: Investigation from a biokinetics modeling perspective. *Bioresource Technology* 2013, 140, 312–318. (Reproduced with Permission from Elsevier, Copyright (2013)).

Xu Wang, Junxin Liu, Nan-Qi Ren, Han-Qing Yu, Duu-Jong Lee, Xuesong Guo. Assessment of multiple sustainability demands for wastewater treatment alternatives: A refined evaluation scheme and case study. *Environmental Science & Technology* 2012, 46(10), 5542–5549. (Reproduced with Permission. Copyright (2012) American Chemical Society).

Xu Wang, Junxin Liu, Nan-Qi Ren, Zuoshan Duan. Environmental profile of typical anaerobic/anoxic/oxic wastewater treatment systems meeting increasingly stringent treatment standards from a life cycle perspective. *Bioresource Technology* 2012, 126, 31–40. (Reproduced with Permission from Elsevier, Copyright (2012)).

Acknowledgements

I joined the Research Center for Eco-Environmental Sciences at the Chinese Academy of Sciences in 2010 and started a three-year study in this excellent institute. Many colleagues and friends have helped me throughout my Ph.D. study. I would like to take this opportunity to express my gratitude to all of them.

Firstly, I would like to express my sincere gratitude to my advisor Prof. Junxin Liu for the continuous support of my Ph.D. study and related research, his patience, motivation, and sharing of his immense knowledge. His guidance helped me in all aspects of this research and writing this thesis. I could not have imagined having a better advisor and mentor.

I would also like to thank a number of leading experts, that is, Profs. Nan-Qi Ren (Harbin Institute of Technology), Jiuhui Qu (Research Center for Eco-Environmental Sciences, Chinese Academy of Sciences), Han-Qing Yu (University of Science and Technology of China), and Duu-Jong Lee (National University of Taiwan), for their insightful comments and encouragement throughout my Ph.D. study and challenging questions that provoked me to view my research from various perspectives.

I also appreciate the stimulating discussions with my fellow labmates. I thank them for the sleepless nights of working together toward deadlines and for all the fun we have had throughout my Ph.D. study.

Last but not least, I would like to thank my parents for supporting me spiritually throughout my Ph.D. study and my life in general.

The research carried out in this thesis was funded by the National Natural Science Foundation of China (Grants 51138009 and 50921064) and Shanghai Tongji Gao Tingyao Environmental Science & Technology Development Foundation (STGEF).

Acknowledgements

Contents

Chapter 1
Introduction

1.1 Study Background

Secondary biological processes, which are currently widely used for the treatment of municipal wastewater, effectively remove substances, such as organic matter, nitrogen, and phosphorus, from wastewater, thereby satisfying certain discharge standards. However, from a modern point of view, the development of such secondary biological processes faces great challenges, mainly based on high energy consumption, excess sludge production, and greenhouse gas (GHG) emissions during the wastewater treatment process and the fact that there is an energy shortage and GHG emissions need to be reduced (Foley et al. 2010; Shahabadi et al. 2009; Verstraete and Vlaeminck 2011). The improvement of the effluent quality is an indicator of wastewater treatment, which has received considerable attention; it is a decisive factor that directly affects investments in wastewater treatment (Roeleveld et al. 1997). Consequently, existing technologies, which aim at ensuring the water quality, remove contaminants from wastewater at the expense of energy and resource consumption without considering the recovery and reclamation of recyclable substances in the wastewater. Although nutrient removal is an important aspect of modern wastewater treatment, existing treatment technologies cannot effectively regenerate nutrients such as nitrogen, phosphorus, and other mineral components. The utilization rate of natural resources therefore exceeds the renewal rate, which leads to resource shortages and creates a vicious cycle (Lampert 2003). In the long term, wastewater treatment processes with high energy consumption and considerable environmental impact will lose competitiveness.

As a matter of fact, the organic matter in wastewater and excess sludge produced by biological treatment units are recyclable resources (Cornel and Schaum 2009; McCarty et al. 2011). The effluent quality should not be the only indicator of the sustainable management of wastewater treatment. In addition to the protection of water resources, the sustainable development of wastewater treatment processes

X. Wang, *Energy Consumption, Chemical Use and Carbon Footprints of Wastewater Treatment Alternatives*, Springer Theses,
https://doi.org/10.1007/978-981-13-5983-5_1

should also focus on other resources in the long term, especially energy and nutrients. Sludge is mainly composed of microbes and organic material. Microbial cells are also organic matter, which can be recovered through methanogenesis during the anaerobic fermentation of sludge, biological hydrogen production, and microbial fuel cells (Juang et al. 2011; Logan et al. 2006; Ren et al. 2011; Show et al. 2011; Sun et al. 2008). Nitrogen and phosphorus in wastewater are also usable resources. Nutrients and minerals have become growingly important resources since the 1960s. Yet, mineral mining consumes large amounts of energy and causes pollution. Furthermore, there is an increasing shortage of high-grade phosphate, potash, and sulfur rocks. In particular, phosphorus is a nonrenewable and irreplaceable resource that mainly exists in the form of natural phosphate ores such as phosphate rock, struvite, and animal fossils. It mainly moves on a one-way path through the biosphere and proven phosphorus reserves are expected to be exhausted in 100 years (Cordell et al. 2009). Some researchers conducted studies on the phosphorus recovery from supernatant generated during anaerobic sludge fermentation, while others proved the feasibility of phosphorus recovery via struvite (Peters and Rowley 2009).

As discussed above, the reduction of the energy consumption and GHG emissions and the recovery of resources are inevitable for the sustainable management of wastewater treatment (Davidson et al. 2010). However, recent research mainly focused on one aspect only, which resulted in the lack of (1) understanding the management of multiple objectives of current treatment technologies such as energy conservation, carbon emission reduction, and resource recovery; (2) in-depth knowledge about the mechanisms and regulation methods for the transfer and conversion of different types of substances during wastewater treatment processes; and (3) comprehensive research on technological wastewater treatment systems. The assessment and optimization of municipal wastewater treatment processes aiming at energy conservation, reduced carbon emissions, and resource recovery therefore currently constitute major challenges in the field of wastewater treatment.

1.2 Study Objectives

At present, secondary biological processes for the treatment of municipal wastewater have several shortcomings such as high energy consumption, GHG emissions, and high excess sludge. In the modern world, which faces an increasing shortage of energy and advocacy of energy conservation and low carbon emissions, the development of such processes will have to overcome huge obstacles. The objectives of this study were as follows: (1) elucidate the transfer, conversion, flow, and allocation of carbon sources and nutrients during wastewater treatment; (2) establish an assessment method and optimization model for the multi-objective management of wastewater treatment processes aiming at energy conservation, carbon emission reduction, and resource recovery; (3) propose wastewater treatment processes that can achieve high efficiency, energy conservation, reduced carbon emissions, and resource recovery to attain the objectives of reducing the energy consumption, reducing contaminant emissions, and enhancing resource recycling; and (4) establish a resource-saving and environmentally friendly technological system that substantially improves resource and energy utilization rates.

1.3 Study Content

In view of the deficiencies of existing methods for the assessment of wastewater treatment technologies, a new technological system for the assessment of multi-objective management of municipal wastewater treatment aiming at energy conservation, reduced carbon emissions, and resource recovery was developed in this study. Based on the comprehensive assessment of existing wastewater treatment processes, a novel model is proposed for the targeted regulation of wastewater treatment to provide a scientific basis and technical support for the research and development of new technologies that can achieve energy conservation and emission reduction, cascade utilization of organic carbon in wastewater, and nutrient recycling.

This study includes the following aspects:

(1) Life cycle inventory analysis of typical wastewater treatment chains (Chap. 3): Life cycle inventory analysis was employed to analyze the environmental impacts of typical municipal wastewater treatment technologies and identify environmental burdens and key links during the operation of different combinations of typical technologies.
(2) A refined assessment methodology for wastewater treatment alternatives (Chap. 4): Based on a literature review, the concept of multi-objective management of municipal wastewater treatment aiming at energy conservation, reduced carbon emissions, and contaminant recycling is proposed. Based on an environmental impact inventory analysis developed in a previous study, the principles of environmental footprint minimization and resource recovery maximization were followed to construct a technological system for the assessment of multi-objective management of municipal wastewater treatment from the aspects of boundary definition, data acquisition, data processing, and comprehensive analysis.
(3) Determination of the weighting element of assessment indicators (Chap. 5): Based on the technological system described in Chap. 4, a global time series database was used to define a weighting system for the multi-objective management assessment system. By using statistical principles, the irreplaceability of each indicator of the indicator system constructed in Chap. 4 was tested to further optimize the indicator system. Subsequently, a weight prediction model based on historical data was built in preparation for the scenario analysis of multi-objective management of municipal wastewater treatment.
(4) Scenario analysis for the multi-objective management of municipal wastewater (Chap. 6): By focusing on the weights of various indicators, scenario analysis and predictions for multi-objective management of wastewater treatment were

performed based on weighting considerations. Reasonable management methods for future municipal wastewater treatment processes were also explored to provide strong scientific support for decision-makers.

(5) Preliminary exploration of sustainability solutions for wastewater management services (Chap. 7): Based on the scenario analysis of multi-objective management of municipal wastewater treatment, the impact of organic carbon on the energy consumption and resource recovery potential during the wastewater treatment process was analyzed using common types of organic carbon in wastewater and whole-process simulation technology. Subsequently, the organic carbon transfer and conversion were used to preliminarily determine the potential regulation of the wastewater treatment process based on multi-objective management.

References

Cordell, D., Drangert, J. O., & White, S. (2009). The story of phosphorus: Global food security and food for thought. *Global Environmental Change-Human and Policy Dimensions, 19*(2), 292–305.

Cornel, P., & Schaum, C. (2009). Phosphorus recovery from wastewater: Needs, technologies and costs. *Water Science and Technology, 59*(6), 1069–1076.

Davidson, C. I., Hendrickson, C. T., Matthews, H. S., Bridges, M. W., Allen, D. T., Murphy, C. F., et al. (2010). Preparing future engineers for challenges of the 21st century: Sustainable engineering. *Journal of Cleaner Production, 18*(7), 698–701.

Foley, J., de Haas, D., Yuan, Z. G., & Lant, P. (2010). Nitrous oxide generation in full-scale biological nutrient removal wastewater treatment plants. *Water Research, 44*(3), 831–844.

Juang, C. P., Whang, L. M., & Cheng, H. H. (2011). Evaluation of bioenergy recovery processes treating organic residues from ethanol fermentation process. *Bioresource Technology, 102*(9), 5394–5399.

Lampert, C. (2003). Selected requirements on a sustainable nutrient management. *Water Science and Technology, 48*(1), 147–154.

Logan, B. E., Hamelers, B., Rozendal, R. A., Schrorder, U., Keller, J., Freguia, S., et al. (2006). Microbial fuel cells: Methodology and technology. *Environmental Science and Technology, 40*(17), 5181–5192.

McCarty, P. L., Bae, J., & Kim, J. (2011). Domestic wastewater treatment as a net energy producer-can this be achieved? *Environmental Science and Technology, 45*(17), 7100–7106.

Peters, G. M., & Rowley, H. V. (2009). Environmental comparison of biosolids management systems using life cycle assessment. *Environmental Science and Technology, 43*(8), 2674–2679.

Ren, N. Q., Guo, W. Q., Liu, B. F., Cao, G. L., & Ding, J. (2011). Biological hydrogen production by dark fermentation: Challenges and prospects towards scaled-up production. *Current Opinion in Biotechnology, 22*(3), 365–370.

Roeleveld, P. J., Klapwijk, A., Eggels, P. G., Rulkens, W. H., & vanStarkenburg, W. (1997). Sustainability of municipal wastewater treatment. *Water Science and Technology, 35*(10), 221–228.

Shahabadi, M. B., Yerushalmi, L., & Haghighat, F. (2009). Impact of process design on greenhouse gas (GHG) generation by wastewater treatment plants. *Water Research, 43*(10), 2679–2687.

Show, K. Y., Lee, D. J., & Chang, J. S. (2011). Bioreactor and process design for biohydrogen production. *Bioresource Technology, 102*(18), 8524–8533.

Sun, M., Sheng, G. P., Zhang, L., Xia, C. R., Mu, Z. X., Liu, X. W., et al. (2008). An MEC-MR-coupled system for biohydrogen production from acetate. *Environmental Science and Technology, 42*(21), 8095–8100.

Verstraete, W., & Vlaeminck, S. E. (2011). ZeroWasteWater: Short-cycling of wastewater resources for sustainable cities of the future. *International Journal of Sustainable Development and World Ecology, 18*(3), 253–264.

Chapter 2
Systematic Literature Review

2.1 Introduction

Secondary biological processes are currently widely used in the treatment of municipal wastewater, and they effectively remove substances (such as organic matter (OM), nitrogen, and phosphorus) from the wastewater, thereby satisfying certain discharge standards (Daigger et al. 2017). However, from the perspective of saving energy and providing environmentally friendly processes, the development of these secondary biological processes faces huge obstacles because they have a high energy consumption and produce excess sludge and greenhouse gas (GHG) emissions during the wastewater treatment process (Curtis 2010; Foley et al. 2010a; Shahabadi et al. 2009). Existing technologies aim to ensure water quality; however, they remove contaminants from wastewater at the expense of energy and resource consumption and do not consider the recovery and reclamation of recyclable substances within the wastewater (Guest et al. 2009). Therefore, from the perspective of long-term wastewater treatment development, treatment processes that involve high energy consumption and a considerable negative environmental impact are less attractive than other process and will be unable to compete with their implementation.

The OM in wastewater and excess sludge produced by biological treatment units are recyclable resources (Cornel and Schaum 2009; McCarty et al. 2011; Verstraete and Vlaeminck 2011). Sludge is mainly composed of microbes and organic material, and microbial cells, which are also OM, can be recovered through methanogenesis during the anaerobic fermentation of sludge, or by sludge composting for agricultural use. However, when nitrogen and phosphorus are biologically removed during the wastewater treatment process, large carbon sources are required; therefore, the OM in wastewater is almost entirely consumed. In addition, sludge has a low OM content in general, which leads to a low energy recovery rate and difficulties in balancing calorific values (McCarty et al. 2011). Nevertheless, nitrogen and phosphorus in wastewater are also usable resources, and phosphorus (in particular) is a nonrenewable and continuously depleted resource. There is currently a lack of emphasis on

X. Wang, *Energy Consumption, Chemical Use and Carbon Footprints of Wastewater Treatment Alternatives*, Springer Theses,
https://doi.org/10.1007/978-981-13-5983-5_2

current treatment processes on the recovery of nitrogen and phosphorus in wastewater, and there is a lack of associated effective technologies. If the recovery rates of nitrogen and phosphorus were increased, the current biological removal processes for nitrogen and phosphorus would require less organic carbon sources. In addition, sludge properties could be improved, which would elevate the resource recovery potential of sludge.

2.2 Research on Municipal Wastewater Treatment Technologies Aimed at Energy Conservation, Reduced Carbon Emissions, and Resource Recovery

2.2.1 Progress in Research on Energy Consumption and Efficiency of Municipal Wastewater Treatment Processes

With the rapid development of modern technologies used in environmental testing and analysis in recent decades, wastewater treatment and purification mechanisms have been thoroughly analyzed. China's municipal wastewater treatment processes have undergone rapid development and enabled increasingly stringent wastewater discharge standards to be satisfied. However, stringent emission standards require an increase in the removal rates of substances such as OM, nitrogen, and phosphorus in wastewater. To achieve the high-speed and large-volume microbial decomposition of OM, increased oxygen supply is required for aeration. Alternatively, new treatment units could be added to the foundation of original treatment processes to meet higher treatment demands. However, such measures require higher energy input or resource consumption, which clearly contradicts the modern concepts of sustainable energy and resource utilization. There is currently a relative lack of in-depth studies on the energy consumption and efficiency of municipal wastewater treatment plants (WWTPs) in China. Instead, current research focuses on the investigation and preliminary analysis of energy consumption through conducting surveys and assessments of energy consumption in addition to operation cost analyses (Hao et al. 2015).

In 1984, several researchers in China used statistic surveys to investigate and estimate electricity consumption in the respective units of primary and secondary biological processes conducted in typical municipal WWTPs (Yang 1984), and results showed that electricity consumption was 0.072 and 0.266 kWh/m^3, respectively. The wastewater and sludge lifting processes and the supply of oxygen for aeration in biological treatment units and sludge digestion were found to be the primary energy-consuming units of the plants. In particular, the biological treatment and sludge treatment units were the major energy consumers at WWTPs, accounting for more than 60% of the total energy consumption. Mover, Yang et al. conducted statistical analysis and quantitative pattern analysis of the annual energy consumption and relevant influencing factors of approximately 600 municipal WWTPs in China (Yang et al. 2008). The results showed that the oxidation ditch, sequencing batch reactor (SBR), and anaerobic/anoxic/oxic (A^2/O) processes, which were the main secondary biological processes used in China, accounted for 32.1, 20.2, and 14.7% of the energy consumption at China's secondary WWTPs, respectively. The average electricity consumption of China's municipal WWTPs was estimated to be around 0.29 kWh/m^3, with the energy consumption of more than 82% of the plants exceeding 0.440 kWh/m^3, which meant that the energy consumption of WWTPs in China might be higher than that of developed countries. Municipal WWTPs are energy-intensive, and the high energy consumption hinders their sustainable operations and long-term development

European countries and the USA both conduct research on energy consumption during municipal wastewater treatment processes. The USA possesses the largest number of municipal WWTPs in the world. However, although their energy consumption is lower than that of China, the energy-intensive nature of wastewater treatment technologies has led to an alarming and continuous increase in the energy budget of WWTPs. Therefore, the U.S. Environmental Protection Agency and Department of Energy have placed great importance on energy consumption and efficiency when planning, designing, and operating water treatment plants. Wesner et al. (Wesner and Culp 1978) were commissioned by the US Environmental Protection Agency to conduct a survey on the energy demands of process units used in municipal WWTPs in the USA. The comprehensive and detailed survey covered almost all US municipal wastewater treatment processes. In addition, a detailed theoretical analysis of wastewater reuse and energy recovery in the respective WWTPs was also conducted, and the total energy demand and recyclable energy of WWTPs with different sizes and different treatment processes were predicted. The survey results reported by Wesner et al. (Middlebrooks et al. 1952) presented a comparative analysis of energy consumption relative to various combinations of wastewater treatment processes conducted in different cities, based on typical influent/effluent quality standards and operating conditions. The results of the study showed that various combinations of wastewater treatment processes that were centered on land treatment had the lowest energy consumption when topographic, land coverage, and groundwater conditions were suitable. In contrast, physical and chemical treatment processes had the highest energy consumption among conventional wastewater treatment methods. Owen (1982) asserted in his monograph that it is a prerequisite to understand the key factors influencing the energy consumption of wastewater treatment if energy conservation and consumption reduction measures are to be implemented. In addition, the relationships between energy utilization during wastewater treatment processes and energy consumption in other industrial sectors should be simultaneously determined. Furthermore, an increased proportion of research conducted by both scientists and engineering personnel should focus on high-efficiency and energy-saving wastewater treatment technologies, with particular emphasis on energy-intensive links in wastewater treatment processes.

2.2.2 Municipal Wastewater Treatment Technologies Aimed at Energy Conservation and Reduced Carbon Emissions

With the severe challenges relating to global energy shortages and climate change, various industries have taken action to conserve energy, cut down emissions, and reduce carbon footprints, and the global water research field is no exception (Poulsen and Hansen 2009; Rosso and Bolzonella 2009; Rosso and Stenstrom 2008). In the past 10 years, many studies have been conducted on energy conservation, emissions reduction, and GHG emissions control in wastewater treatment systems, and significant research progress has been made (Insam and Wett 2008; Plosz et al. 2009; Shahabadi et al. 2010; Smith 2009).

In this respect, the energy consumption characteristics and energy conservation technologies of wastewater treatment processes have been widely discussed. In a study by Jin et al., WWTPs were divided based on function and energy consumption characteristics into three major units: pretreatment, biochemical treatment, and sludge treatment. Through this analysis, a customized method for analyzing WWTP energy consumption was established and can be used to select feasible energy conservation paths (Jin et al. 2009). Yang et al. used computational fluid dynamics (CFD) tools to simulate and optimize the three-dimensional flow field of oxidation ditches, to improve the mass transfer efficiency of aeration within reactors and reduce electricity consumption (Yang et al. 2011). Several other researchers have also conducted studies on the simulation and optimization of aeration in aerobic tanks for energy conservation (Gresch et al. 2011; Rosso et al. 2008).

With respect to GHG generation and emissions reduction in wastewater treatment processes, Fukushima et al. (2008) preliminarily investigated the GHG emissions of Taiwan's environmental protection industrial parks. The results showed that the CO_2 equivalent emissions from wastewater treatment in Taiwan in 2004 accounted for 21.5% of the total emissions, thus making wastewater treatment one of the largest sources of emissions among all of these industries. Bruins et al. (1995) studied the GHG emissions of WWTPs in the Netherlands in 1987 and reported CO_2 emissions of 1.2 kg CO_2/kg COD. The CO_2 emissions of the various units of wastewater treatment processes were also analyzed, and it was found that aerobic biological treatment units accounted for approximately 75% of total CO_2 emissions. In some countries, the emission factors of CH_4 and N_2O have been determined by investigating the mechanisms of GHG emissions in municipal wastewater treatment processes (MFE 2003; Thomsen and Lyck 2005). Preliminary results have shown that the magnitudes of emission factors are related to wastewater characteristics and the type of treatment process employed. Cakir and Stenstrom (2005) compared GHG generation in aerobic and anaerobic wastewater treatment processes and concluded that less GHGs were produced by the aerobic treatment process when the biochemical oxygen demand

(BOD) of biological wastewater was less than 300 mg/L. In addition, the researchers recommended that aerobic post-treatment of the effluents of anaerobic treatment units was should be conducted to effectively oxidize CH_4 dissolved in water and prevent the escape of CH_4 into the atmosphere. Wong et al. (2009) studied the carbon balance of anaerobic granular sludge treatment systems. As anaerobic granular sludge is characterized by high load resistance and short hydraulic retention time, the anaerobic granular sludge treatment process produces less carbon emissions than other anaerobic treatment technologies, and if this technology was applied to typical anaerobic digesters, CH_4 emissions would be reduced by 46.42 tons of CO_2 equivalent per year. Other studies have reported that wastewater treatment processes produce relatively low amounts of N_2O emissions (Foley et al. 2010b; Kampschreur et al. 2009), however, they constitute a relatively large emission source in the water industry, accounting for approximately 26% of total emissions. A lower dissolved oxygen concentration or higher nitrate concentration in the nitrification and denitrification stages, or a lower COD/N ratio in the denitrification stage, leads directly lead to the release of N_2O; however, it is not yet known whether the release is caused by nitrifying or denitrifying bacteria. On the basis of research conducted on microbial fuel cells (MFCs), Wang et al. (2010) pioneered the addition of Microcystis in an electrode chamber to construct microbial carbon capture cells (MCCs). The MCCs were proven to be able to utilize the OM in wastewater to generate electricity and capture the CO_2 generated during the degradation of OM, thereby providing a novel concept for energy conservation and emissions reduction in wastewater treatment.

2.2.3 Municipal Wastewater Treatment Technologies Aimed at Resource Recovery

In recent years, it has become gradually apparent that the sustainable development of water resource and wastewater treatment systems is important. If the research, development, and wide application of municipal wastewater treatment technologies can enable the full recovery of energy and nutrients from wastewater under the premise that water quality is maintained, these technologies will be of profound significance to sustainable development.

The OM in wastewater is a potential source of green biomass energy. Each kilogram of OM (in terms of COD) contains approximately 14 MJ of metabolic heat (Rabaey et al. 2010). However, energy has been consumed in the removal of such OM and its recovery has not yet been considered. Researchers reported that the energy contained in wastewater processed in an ordinary WWTP in Toronto, Canada, was 9.3 times higher than the energy consumed for wastewater treatment (Shizas and Bagley 2004). According to another study (Logan 2004), wastewater currently produced from animal and food processing in the USA contains approximately 17 GW of energy, which is roughly equivalent to the energy currently consumed by all water infrastructures in the USA. Therefore, if this energy source can be recycled, it would be possible to achieve energy self-sufficiency in water infrastructures. In recent years, methods for the production of energy and electricity from OM in wastewater and sludge, including methanogenesis via anaerobic fermentation, biological hydrogen production, and MFCs, have become popular focuses of domestic and international research (Juang et al. 2011; Logan et al. 2006; Ren et al. 2011; Show et al. 2011; Sun et al. 2008). The study by Rullens et al. proposed a concept and process for maximizing energy production from biomass in municipal wastewater (Rullkens et al. 2009). Verstraete et al. reviewed the technologies used to maximize the recovery of municipal wastewater resources and energy sources and proposed a method for the concentration of domestic sewage to achieve the utilization of OM as an energy source (Verstraete and Vlaeminck 2011).

Nutrients and minerals have become increasingly important resources since the 1960s, but mineral mining consumes large amounts of energy and causes pollution. Furthermore, there is an increasing shortage of high-grade phosphate, potash, and sulfur rocks. In particular, phosphorus is a nonrenewable and irreplaceable resource that mainly exists in the form of natural phosphate ore, such as phosphate rock, struvite, and animal fossils. It mainly moves on a one-way path through the biosphere, and proven phosphorus reserves are expected to be exhausted in 100 years. (Cordell et al. 2009) In addition to OM, nitrogen and phosphorus are also usable resources in wastewater, and another advantage of recovering phosphorus during the wastewater treatment process is the concurrent removal of phosphorus from wastewater. In recent years, the feasibility of combined phosphorus removal and recovery has been proven by relevant research, particularly by studies on phosphorus recovery from sludge reflux and sludge liquor, such as the recovery of struvite from supernatant in anaerobic digestion (Bauer et al. 2007; Le Corre et al. 2009; Massey et al. 2009). In addition, a study showed that when phosphorus-rich sludge reflux or liquor was fed into a crystallizer and calcium or magnesium salts were added, phosphorus in the liquid phase was precipitated in the form of calcium phosphate or magnesium ammonium phosphate (Yuan et al. 2012). In another study, it was reported that approximately 90% of the phosphorus in wastewater leaves the wastewater treatment system along with sludge discharge (Metcalf and Eddy 2003). Therefore, the recovery of phosphorus from sludge is one of the key steps in the promotion of the natural phosphorus cycle. Currently, wet chemical and thermochemical processes are the technologies commonly used for phosphorus recovery from sludge (Petzet and Cornel 2011; Thygesen et al. 2011). In wet chemical methods, the phosphorus in sludge is first dissolved by adding an acid or alkali, and the phosphorus in the solution can then be separated by precipitation, ion exchange, nanofiltration, or liquid–liquid extraction/separation. In thermochemical methods, to increase the phosphorus recovery rate, the sludge is combusted and specific thermochemical treatment is subsequently applied to remove heavy metals from the incineration ash of the sludge.

2.3 Comprehensive Assessments of Municipal Wastewater Treatment Processes

Since the early 1980s, assessments of municipal wastewater treatment technologies have been conducted globally. Many in-depth studies have been conducted using various decision-making methods and multi-objectives, and over time the focus has gradually shifted from assessments based on a single economic objective to assessments based on multiple objectives, such as economic, technological, and environmental objectives. This transition may now enable the use of municipal wastewater treatment technologies to achieve the scientific allocation of energy and resources while optimizing social, economic, and environmental benefits. The two most representative types of assessment methods are as follows: (1) conventional assessment methods in which the assurance of water quality is set as the objective and wastewater treatment efficiency, effluent quality, and techno-economic feasibility are used as assessment criteria; and (2) assessment methods based on the environmental impacts of wastewater treatment processes.

2.3.1 Assessment of Municipal Wastewater Treatment Based on the Assurance of Water Quality

Environmental pollution issues are becoming increasingly important, and with additional heightened concerns about health and safety and production environments, national and local regulations on wastewater discharge have become increasingly stringent. Therefore, greater demands have been placed on the efficiency of wastewater treatment and the degree to which it is treated. Currently, and for many years, techno-economic analysis methods based on the assurance of water quality have been the main methods used to assess municipal wastewater treatment processes or schemes (Aydiner et al. 2014; Kobya et al. 2007; Xin et al. 2016). Under the premise that water quality meets relevant standards, such methods involve consideration of the benefits and expenses involved in the wastewater treatment process (or scheme) from a national economy point of view and use the principle of the rational allocation of resources and analyze the economic rationality of the assessed subject.

One indicator of wastewater treatment that has been the focus of much attention is the improvement of effluent quality; it is also a decisive factor that directly affects investments in wastewater treatment. Consequently, existing processes, which aim to assure water quality, remove contaminants from wastewater at the expense of energy and resource consumption and do not consider other possible environmental impacts that may arise from wastewater treatment.

2.3.2 Assessment of Municipal Wastewater Treatment Based on Environmental Impacts

Over the past two decades, municipal wastewater treatment technologies have undergone continuous development and become increasingly innovative, and municipal wastewater treatment systems focusing on secondary biological processes have become more widely used. Municipal wastewater treatment reduces water pollution and plays an active role in the reduction in eutrophication in surface water environments, and conventional methods used to assess and select wastewater treatment processes commonly employ a techno-economic analysis based on the assurance of water quality (see Sect. 2.3.1). However, these methods have minimal consideration for the environmental impacts of different wastewater treatment schemes. At present, the concept of sustainable development is a deeply rooted concept, and the environmental impact of effluents from WWTPs and entire wastewater and sludge treatment processes are receiving increasing attention. Therefore, under the premise that water quality standards are satisfied, it has become necessary to conduct a comprehensive assessment of wastewater treatments using an environmental impact analysis.

The life cycle assessment (LCA) is a technique used to assess the environmental load associated with all stages of the life cycle of a product, process, or activity, from the stage of raw material extraction through processing, manufacture, transportation, sales, use, recovery, maintenance, recycling, and disposal (Dennison et al. 1998). The LCA, which is currently widely applied internationally, can be effectively used to assess the entire production process of a product, or it can be employed as a tool for environmental management. More than a decade ago, LCA methodology was used to assess entire wastewater treatment processes, and many research achievements have since been published in journals globally. The LCA has thus become the mainstream tool for providing an analysis of the environmental impacts of municipal wastewater treatment processes (Emmerson et al. 1995; Hospido et al. 2004; Lundin et al. 2000; Zhang and Wilson 2000).

A previous study (Papa et al. 2016) applied the LCA in a comparative case study focusing on rethinking wastewater treatment trains based on their impacts and benefits on human health, with the aim of assessing the entire process at each wastewater treatment facility (from facility design through materials and energy acquisition, construction, operation management, scrapping, and demolition). By linking technical, economic, social, and psychological factors with environmental protection, the advantages and disadvantages of the two treatment methods were comprehensively determined. Zhang and Wang employed the LCA as a tool to identify and quantitatively analyze the resource consumption, energy consumption, and environmental impacts of the entire life cycle of the wastewater treatment process at the Beishiqiao Sewage Treatment Center in Xi'an, China (Zhang and Wang 2009). The results showed that advanced treatment of wastewater provides huge environmental benefits, although it is energy-intensive. In addition, Pasqualino et al. (2009) employed the LCA as a tool for devising optimal operation measures for municipal WWTPs. The treatment processes of the existing municipal WWTPs and alternative optimized processes were analyzed and compared, and recommendations for the optimization, operation, upgrade, and reconstruction of WWTPs were provided on the basis of study results. In the past two years, researchers have performed comprehensive assessments to report the link between wastewater treatment processes and energy consumption and climate change impacts (Akerman 2010; Bishop 2010; Crawford 2010; Kenway 2010).

Introduction of the LCA method in the past decade has enabled an in-depth understanding of the environmental impacts of wastewater treatment processes. However, conducting LCA is time-consuming and labor-intensive, and the LCA also has certain limitations when applied in some environmental disciplines, such as wastewater and sludge treatment processes (Renou et al. 2008; Roux et al. 2010; Santero et al. 2011; Tangsubkul et al. 2005). For example, the recovery of usable substances in wastewater and sludge has not yet been considered. Furthermore, in the two assessment systems described above (i.e., the techno-economic and LCA-based approaches), there is a lack of comprehensiveness with respect to the assessment scope, and assessment data are mainly obtained through field investigations, personal interviews, or internal reports of the plants. Therefore, there is a risk that data uncertainty caused by subjective factors may also affect the reliability of the assessment results.

2.4 Mathematical Models and Simulation Platforms for Municipal Wastewater Treatment

With increasing demands for optimal water quality and the rise in corresponding water quality standards, it has become a priority to construct and upgrade various types of municipal WWTPs. Mathematical wastewater treatment models are tools that can be used to guide and assist construction and renovation projects, and they have been gradually receiving attention and have been tentatively applied to actual production (Langergraber et al. 2004). Research on mathematical models of wastewater treatment has progressed through a development process involving three stages: simple fitting of experimental data, employment of typical dynamic microbial growth models, and modeling through dynamic analysis, exploration, and identification of wastewater biological treatment processes based on process characteristics. In this respect, the main objective has shifted from guiding the design of wastewater treatment processes to investigating the dynamic processes involved in wastewater treatment, with the aim of achieving high operational efficiency and low energy consumption. On the basis of such models, corresponding commercial wastewater treatment software has now been developed (Gernaey et al. 2004).

Establishing a comprehensive and practical mathematical model is not only important for the design, operation, and management of biological treatment processes for industrial wastewater, but also serves as a key reference for the design of control strategies (Gernaey et al. 2004). Currently, mathematical models widely used for wastewater treatment are mostly derived from the activated sludge process. The basic principle is as follows: the Monod equation representing cell growth dynamics is used together with reactor theories, microbiological theories, and fluid mechanics theories of the chemical engineering field, with the aim of quantitatively describing the relationships between matrix degradation, microbial growth, and contaminant concentration distribution (with various parameters) to enable model construction (Hauduc et al. 2013). Modeling technologies have developed rapidly and have become increasingly mature, and current wastewater treatment simulation platforms can track the variation data of any model component or state variable of the different process units of a WWTP. As such, this has become another feasible way of obtaining wastewater treatment assessment data, which potentially compensates for the inadequacies of conventional data acquisition methods (see related description in Sect. 2.3.2).

2.4.1 Application of Mathematical Models in Wastewater Treatment Processes

The application of mathematical simulation technologies has received widespread attention globally. Mathematical simulation technologies have evolved from being purely theoretical to being widely applied in engineering and research. Mathematical models of biological wastewater treatment, including models for the activated sludge process, biofilm process, removal of OM and nutrients, externally described black box models, and internally linked metabolic models, have become fundamentally robust after 20–30 years of development (Gernaey et al. 2004; Hauduc et al. 2013). With the rapid development of computer technology in the past two decades, the application of mathematical models has become a reality; they can be established based on the results of fundamental research (e.g., microbiology) (Kowalchuk and Stephen 2001; Ni and Yu 2010, 2012; Nielsen et al. 2012; Seviour et al. 2003) and also directly used in small-scale tests for the correction of dynamic parameters (Donoso-Bravo et al. 2011; Hu et al. 2012; McLean and McAuley 2012). With the support of mathematical simulations, the results of small-scale tests can be directly extended to engineering design, thereby eliminating the need for pilot studies (Fenu et al. 2010; Gohle et al. 1996). Simulations can also play an important role in optimizing the operations of actual processes (Belia et al. 2009; Hauduc et al. 2009; Jimenez-Hornero et al. 2009; Lee et al. 2006; Liu et al. 2010; Patsios and Karabelas 2010; Sweeney and Kabouris 2004; Yang et al. 2010).

2.4.2 Mathematical Simulation Platforms for Wastewater Treatment Processes

Mathematical simulation platforms for wastewater treatment processes are dynamic model platforms that show dynamic information about the wastewater treatment process and which can be packaged and integrated. Such computer-based experimental platforms are powerful mathematical tools that integrate design, planning, process analysis, and operation guidance, and they can also be used to monitor dynamic information about wastewater treatment processes. Such platforms have thus been used in the field of wastewater treatment for many years, and the review of Li highlighted the existence of two main mathematical simulation platforms currently used in wastewater treatment, namely open simulation platforms and specialized wastewater treatment simulation software (Li et al. 2015).

Open numerical simulation platforms are mainly developed based on common programming languages (such as C/C++ and FORTRAN) or scientific and engineering calculation software (such as MATLAB/SIMULINK). Unlike specialized simulation software for wastewater treatment, open simulation platforms provide highly flexible modeling and computing environment. On such platforms, it is possible to simulate wastewater treatment processes according to specific mathematical descriptions. However, although the modeling process is independent, it is more time-consuming compared with modeling using specialized wastewater treatment simulation software. In addition, modelers need a solid foundation in mathematics and programming to provide an understanding of model uncertainties and enable the rapid identification and resolution of problems, such as iterative errors. Open simulation platforms have thus been developed for modelers who are experienced in wastewater treatment modeling, and such modelers are required to fully understand each code of the model. Currently, the open simulation platform widely used in the field of wastewater treatment is AQUASIM (Reichert 1994), which was developed by the Swiss Federal Institute of Aquatic Science and Technology (EAWAG), and is an open modeling platform based on a graphical interface system. Modelers can build individual models of wastewater treatment processes based on actual needs. Moreover, AQUASIM is built on a certain number of predefined hierarchical models, and modelers can select the corresponding hierarchical model and numerical algorithm based on actual needs (Boltz et al. 2011; Fall and Loaiza-Navia 2007; Larrea et al. 2007; Salem et al. 2002; Shanahan and Semmens 2004).

Specialized wastewater treatment simulation software often integrates a predefined model library of wastewater treatment process units, such as the well-developed activated sludge model (ASM) and anaerobic digestion model (ADM). Modelers can add and connect process unit modules based on actual process requirements and can also modify model parameters through module properties. At present, most commercial wastewater treatment simulation software has been developed outside of China, and BioWin, GPS-X, SSSP, EFOR, SASS, JASS, WEST, STOAT, and SIMBA are used in a wide range of applications (Makinia 2010). These simulation software packages enable modelers to quickly build process modules for wastewater treatment. Among them, BioWin is the only software that uses self-developed models, while other simulation software employs mathematical models developed by the International Water Association (IWA) (Envirosim 2007). BioWin is mainly based on the Barker and Dold model that integrates various activated sludge dynamic models (Barker and Dold 1997; Dold et al. 1980). It can simulate the water and sludge lines of an entire WWTP (including all physical, chemical, and biological process units) and track the variations of any model component, state variable of the plant in different process units, and of more than 60 physical, chemical, and biological reactions (including the activated sludge process, anaerobic digestion, side stream treatment processes, gas transfer, and chemical precipitation reactions) that occur in these model components (Cherchi et al. 2009; Eldyasti et al. 2012; Hafez et al. 2010; Jones et al. 2009; Liwarska-Bizukojc and Biernacki 2010; Liwarska-Bizukojc et al. 2011; Phillips et al. 2009; Simsek et al. 2012).

2.5 Weighting Systems Commonly Used in Comprehensive Assessment Systems

Weighting can also be known as the weighting factor, and it refers to the quantitative assignment of the importance of an object or indicator (Guinee 2002; Horne et al. 2009; Schmidt and Sullivan 2002). In comprehensive assessment systems with multiple indicators, weighting is used to indicate the relative importance of each assessment indicator, and it thus reflects the importance of each assessment indicator with respect to the assessment goal. Weighting mainly has the following characteristics (Bengtsson and Steen 2000): (1) it reflects the purpose and value orientation of the assessment and highlights the key points; (2) it affects the assessment conclusions, as the weights of various assessment indicators are mutually influential and constrained: When the weight of an assessment indicator is set to a larger value, the weights of other indicators will be smaller, and the variation of weights may also alter the assessment result, thereby resulting in changes in predictions or decisions; and (3) as the assessment goal or concept changes, the weight of the same assessment indicator also needs to be changed: weights should thus be appropriately determined and adjusted based on the actual relative importance of the assessment intentions and indicators (the contributions of the assessment indicators to the comprehensive assessment). In addition, to enhance the comparability between assessment indicators, weights should reflect the respective importance of each considered factor within the comprehensive assessment.

A scientific and reasonable comprehensive assessment of wastewater treatment usually requires the establishment of a scientific and robust assessment indicator system, and it is necessary for the assessment process that the weight of each assessment indicator can be scientifically and objectively reflected. Failing to set weights or setting biased weights may result in the following: uncertainty in the assessment results, logical errors in the process, misleading judgments, and difficulty conducting actual calculations. The determination of weighting systems has always been controversial, and the weights obtained by different methods vary significantly. This is mainly attributed to two reasons: (1) different experts or related personnel involved in the assessment attach different levels of importance to each assessment indicator; and (2) variations in the amount and reliability of information reflected by each indicator: the former mainly reflects the subjectivity of weights, while the latter mainly reflects the objectivity of weights (i.e., the objective influence of different indicators). Therefore, it is essential to use a scientific method to ensure the science and rationality of weighting systems.

In general, methods for determining weighting systems can be classified into two major categories: subjective value assignments and objective value assignments (Finnveden et al. 2009; Soares et al. 2006).

2.5.1 Subjective Value Assignment

In subjective value assignment methods (often referred to as expert-led methods), a number of experts score the importance of each assessment indicator, and the scores of each expert are processed by mathematical methods to obtain the weight of each indicator (Bengtsson and Steen 2000). The most commonly used methods are the expert consultation method and analytic hierarchy process.

(1) Expert consultation method

The most commonly used method in subjective value assignment is the expert consultation method. This method analyzes, judges, and weighs the assessment indicators and assigns weights based on the professional foundation, knowledge reserve, experience accumulation, and value orientation of each expert (Hosftetter et al. 1999). This method generally requires multiple rounds of scoring to perform a statistical analysis on the assessment results of each expert and to test the degrees of concentration, dispersion, and coordination of all data. The initial weight vector of each assessment indicator is determined under certain assumed conditions, and the actual weight vector is obtained after normalization. This method is easy to operate but has many limitations. For example, the selection of experts is in itself subjective, and this can lead to high uncertainty in the results.

(2) Analytic hierarchy process

The analytic hierarchy process is a multi-rule assessment method that is easy to operate (Ong et al. 2001). The assessment is performed based on the inherent logical relationships between complex assessment objects, and the assessment indicators form an orderly hierarchical structure. For each assessment indicator, experts are asked to compare the indicators in pairs. A judgment matrix is constructed according to the predetermined scale values, and the maximum eigenvalue of the matrix and corresponding eigenvector are then calculated. Finally, the feature vector is normalized to obtain the weight vector of each assessment indicator.

2.5.2 *Objective Value Assignment*

Objective value assignment methods are based on the intrinsic relationships between various assessment indicators and enable a direct mathematical calculation of weight values. The initial data are derived from actual data in a matrix, which thus eliminates interference from subjective factors (Bengtsson and Steen 2000). The current commonly used objective value assignment methods include the distance-to-target method and the monetary quantification method.

(1) Distance-to-target method

The distance-to-target method is the most widely applied method of weight determination in environmental impact assessments (Finnveden et al. 2002). The premise is to determine the weight based on the "distance" between the current environmental conditions and an environmental standard. This standard can be a national or regional contaminant discharge standard or the maximum amount of contaminants that cause damage to the environment. However, the "distance" in this method can only reflect severity within an environmental impact category, but it cannot objectively reflect severity among various environmental impact categories. Therefore, in some sense, the distance-to-target method is not considered a weighting method (Finnveden et al. 2002).

(2) Monetary quantification method

The idea of the monetary quantification method is that the severity of assessment indicators can be measured using monetary values. This method is widely used in environmental impact assessments (Finnveden et al. 2006). In the Swedish environmental assessment, the willingness of residents to pay for environmental impacts is used to reflect the weight of each environmental impact category, as proposed by the Organization for Economic Co-operation and Development (OECD). In such assessment systems, market prices are generally used to estimate a willingness to pay, and if market prices cannot be obtained, other indirect methods are used to obtain the values.

2.6 Problems and Prospects

(1) In the past, research and development of municipal wastewater treatment technologies have focused on one advantage and has been dominated by process performance research. However, in recent years, emphasis on the reduction in energy consumption and GHG emissions and the recovery of resources has caused a change in focus with respect to the development of new wastewater treatment technologies. Nevertheless, recent research has mainly focused on a single aspect, which has resulted in a lack of (i) fundamental research on the

transformation of existing processes to future processes that facilitate the reduction in energy consumption and GHG emissions and the recovery of resources, (ii) in-depth knowledge about the mechanisms and regulation methods involved in the transfer and conversion of different types of substances during wastewater treatment processes, and (iii) comprehensive research on different wastewater treatment systems.

(2) Existing comprehensive assessment systems for municipal wastewater treatment mainly assess the treatment efficiency, water quality compliance, and environmental benefits without considering the recovery of recyclable substances in wastewater. In addition, the data sources for the techno-economic and LCA-based approaches described above are influenced by subjective factors to varying degrees, which may lead to a certain magnitude of uncertainty in the assessment results.

(3) The rapid development of mathematical models and simulation platforms for wastewater treatment have largely enabled the integration of models of physical, chemical, and biological treatment units in wastewater treatment processes. These models (or platforms) can simulate entire wastewater and sludge treatment processes and track variations in any model component or state variable within the different process units of an entire WWTP. As such, they provide a feasible method of obtaining data for the assessment of municipal wastewater treatment technologies.

(4) During the comprehensive assessment of municipal wastewater treatment processes, the determination of appropriate weights for the various assessment indicators is the key to obtain scientific assessment results. Commonly used weight assignment methods have their own advantages and disadvantages, and the sole use of a single method may lead to biases in the assigned weights.

(5) In the future, effluent quality should not be the only indicator used to measure the sustainable management of wastewater treatment. In the long term, in addition to the protection of water resources, sustainable development should also focus on other resources, such as energy and nutrients. It is considered that reducing energy consumption, lowering GHG emissions, and recovering resources will be the main focuses of future technologies and processes developed for wastewater treatment. Future trends in wastewater treatment research should use existing technologies and processes to establish new integrated technologies and wastewater treatment processes, with the aim of conserving energy, reducing carbon emissions, and recovering resources. Current systems used in the comprehensive assessment of wastewater treatment are based on water quality assurance, economic benefits, and environmental benefits, and the recovery of usable substances in wastewater and sludge is not adequately considered. Therefore, a scientific, systematic, and comprehensive system for the assessment of wastewater treatment is required. Following a comprehensive assessment of existing technologies and processes, changes should be made to the existing research and development concepts applied to wastewater treatment technologies and processes. Both the theoretical and technical aspects should be considered, and energy conservation, reduction in carbon emissions, and the

recovery of organic nutrient resources should be the main focus. Research on regulating and optimizing the transfer and conversion of different types of substances is thus required, and by combining relevant studies with the research and development of new highly efficient and energy-saving wastewater treatment technologies, a technical system for domestic sewage treatment that meets the multiple objectives of water quality, energy conservation, low carbon emissions, and resource recovery can potentially be developed.

References

Akerman, A. (2010). *Methodology for the assessment of the greenhouse gas impact of wastewater sludge treatment*. Canada: Montreal.

Aydiner, C., Sen, U., Topcu, S., Ekinci, D., Altinay, A. D., Koseoglu-Imer, D. Y., et al. (2014). Techno-economic viability of innovative membrane systems in water and mass recovery from dairy wastewater. *Journal of Membrane Science, 458*, 66–75.

Barker, P. S., & Dold, P. L. (1997). General model for biological nutrient removal activated-sludge systems: Model presentation. *Water Environment Research, 69*(5), 969–984.

Bauer, P. J., Szogi, A. A., & Vanotti, M. B. (2007). Agronomic effectiveness of calcium phosphate recovered from liquid swine manure. *Agronomy Journal, 99*(5), 1352–1356.

Belia, E., Amerlinck, Y., Benedetti, L., Johnson, B., Sin, G., Vanrolleghem, P. A., et al. (2009). Wastewater treatment modelling: Dealing with uncertainties. *Water Science and Technology, 60*(8), 1929–1941.

Bengtsson, M., & Steen, B. (2000). Weighting in LCA—Approaches and applications. *Environmental Progress, 19*(2), 101–109.

Bishop, P. L. (2010). *Water sustainability: The need for enabling research*. USA: Newark.

Boltz, J. P., Morgenroth, E., Brockmann, D., Bott, C., Gellner, W. J., & Vanrolleghem, P. A. (2011). Systematic evaluation of biofilm models for engineering practice: Components and critical assumptions. *Water Science and Technology, 64*(4), 930–944.

Bruins, J. G., Oostergo, H. D., & Brinkhorst, M. A. (1995). In S. Zwerver, R. S. A. vanRompaey, M. T. J.Kok, & M. M. Berk (Eds.), *Climate change research: Evaluation and policy implications, A and B* (pp. 635–638).

Cakir, F. Y., & Stenstrom, M. K. (2005). Greenhouse gas production: A comparison between aerobic and anaerobic wastewater treatment technology. *Water Research, 39*(17), 4197–4203.

Cherchi, C., Onnis-Hayden, A., El-Shawabkeh, I., & Gu, A. Z. (2009). Implication of using different carbon sources for denitrification in wastewater treatments. *Water Environment Research, 81*(8), 788–799.

Cordell, D., Drangert, J. O., & White, S. (2009). The story of phosphorus: Global food security and food for thought. *Global Environmental Change-Human and Policy Dimensions, 19*(2), 292–305.

Cornel, P., & Schaum, C. (2009). Phosphorus recovery from wastewater: Needs, technologies and costs. *Water Science and Technology, 59*(6), 1069–1076.

Crawford, G. V. (2010). *From the best practices to the plant of the future: The WERF optimization program*. Canada: Montreal.

Curtis, T. P. (2010) In: R. Mitchell & J. D.Gu (Eds.), *Environmental microbiology*. Hoboken, NJ: Wiley-Blackwell.

Daigger, G. T., Murthy, S., Love, N. G., & Sandino, J. (2017). Transforming environmental engineering and science education, research, and practice. *Environmental Engineering Science, 34*(1), 42–50.

Dennison, F. J., Azapagic, A., Clift, R., & Colbourne, J. S. (1998). Assessing management options for wastewater treatment works in the context of life cycle assessment. *Water Science and Technology, 38*(11), 23–30.

Dold, P. L., Ekama, G. A., & Marais, G. R. (1980). The activated sludge process. 1. A general model for the activated sludge process. *Progress in Water Technology, 12*(6), 47–77.

Donoso-Bravo, A., Mailier, J., Martin, C., Rodriguez, J., Aceves-Lara, C. A., & Vande Wouwer, A. (2011). Model selection, identification and validation in anaerobic digestion: A review. *Water Research, 45*(17), 5347–5364.

Eldyasti, A., Nakhla, G., & Zhu, J. (2012). Development of a calibration protocol and identification of the most sensitive parameters for the particulate biofilm models used in biological wastewater treatment. *Bioresource Technology, 111,* 111–121.

Emmerson, R. H. C., Morse, G. K., & Lester, J. N. (1995). The life-cycle analysis of small-scale sewage-treatment processes. *Journal of the Chartered Institution of Water and Environmental Management, 9*(3), 317–325.

Envirosim. (2007). *BioWin process simulator*. Envirosim Associates Ltd.

Fall, C., & Loaiza-Navia, J. L. (2007). Design of a tracer test experience and dynamic calibration of the hydraulic model for a full-scale wastewater treatment plant by use of AQUASIM. *Water Environment Research, 79*(8), 893–900.

Fenu, A., Guglielmi, G., Jimenez, J., Sperandio, M., Saroj, D., Lesjean, B., et al. (2010). Activated sludge model (ASM) based modelling of membrane bioreactor (MBR) processes: A critical review with special regard to MBR specificities. *Water Research, 44*(15), 4272–4294.

Finnveden, G., Eldh, P., & Johansson, J. (2006). Weighting in LCA based on ecotaxes—Development of a mid-point method and experiences from case studies. *International Journal of Life Cycle Assessment, 11*(1), 81–88.

Finnveden, G., Hauschild, M. Z., Ekvall, T., Guinee, J., Heijungs, R., Hellweg, S., et al. (2009). Recent developments in Life Cycle Assessment. *Journal of Environmental Management, 91*(1), 1–21.

Finnveden, G., Hofstetter, P., Bare, J., Basson, L., Ciroth, A., Mettier, T., et al. (2002) *Life-cycle impact assessment: Striving towards best practise* (pp. 177–208). Pensacola, FL: SETAC Press.

Foley, J., de Haas, D., Hartley, K., & Lant, P. (2010a). Comprehensive life cycle inventories of alternative wastewater treatment systems. *Water Research, 44*(5), 1654–1666.

Foley, J., de Haas, D., Yuan, Z. G., & Lant, P. (2010b). Nitrous oxide generation in full-scale biological nutrient removal wastewater treatment plants. *Water Research, 44*(3), 831–844.

Fukushima, Y., Liu, P. W. G., Tsai, J. H., Lee, C. F., & Tseng, T. K. (2008). Preliminary investigation of greenhouse gas emissions from the environmental sector in Taiwan. *Journal of the Air and Waste Management Association, 58*(1), 85–94.

Gernaey, K. V., van Loosdrecht, M. C. M., Henze, M., Lind, M., & Jorgensen, S. B. (2004). Activated sludge wastewater treatment plant modelling and simulation: State of the art. *Environmental Modelling and Software, 19*(9), 763–783.

Gohle, F., Finnson, A., & Hultman, B. (1996). Dynamic simulation of sludge blanket movements in a full-scale rectangular sedimentation basin. *Water Science and Technology, 33*(1), 89–99.

Gresch, M., Armbruster, M., Braun, D., & Gujer, W. (2011). Effects of aeration patterns on the flow field in wastewater aeration tanks. *Water Research, 45*(2), 810–818.

Guest, J. S., Skerlos, S. J., Barnard, J. L., Beck, M. B., Daigger, G. T., Hilger, H., et al. (2009). A new planning and design paradigm to achieve sustainable resource recovery from wastewater. *Environmental Science and Technology, 43*(16), 6126–6130.

Guinee, J. B. (2002). *Handbook on life cycle assessment: Operational guide to the ISO standards.* Springer.

Hafez, H., Naggar, M. H. E., & Nakhla, G. (2010). Steady-state and dynamic modeling of biohydrogen production in an integrated biohydrogen reactor clarifier system. *International Journal of Hydrogen Energy, 35*(13), 6634–6645.

Hao, X. D., Liu, R. B., & Huang, X. (2015). Evaluation of the potential for operating carbon neutral WWTPs in China. *Water Research, 87,* 424–431.

Hauduc, H., Gillot, S., Rieger, L., Ohtsuki, T., Shaw, A., Takacs, I., et al. (2009). Activated sludge modelling in practice: An international survey. *Water Science and Technology, 60*(8), 1943–1951.

Hauduc, H., Rieger, L., Oehmen, A., van Loosdrecht, M. C. M., Comeau, Y., Heduit, A., et al. (2013). Critical review of activated sludge modeling: State of process knowledge, modeling concepts, and limitations. *Biotechnology and Bioengineering, 110*(1), 24–46.

Horne, R., Grant, T., & Verghese, K. (2009). *Life Cycle Assessment: Principles, practice and prospects*. VIC, Australia: CSIRO Publishing.

Hosftetter, P., Braunsehwelg, A., & Mettler, T. (1999). The mixing triangle: Correlation and graphical decision support for LCA-based comparisons. *Journal of Industrial Ecology, 3*(4), 97–115.

Hospido, A., Moreira, M. T., Fernandez-Couto, M., & Feijoo, G. (2004). Environmental performance of a municipal wastewater treatment plant. *International Journal of Life Cycle Assessment, 9*(4), 261–271.

Hu, Z., Lee, J. W., Chandran, K., Kim, S., & Khanal, S. K. (2012). Nitrous Oxide (N_2O) emission from aquaculture: A review. *Environmental Science and Technology, 46*(12), 6470–6480.

Insam, H., & Wett, B. (2008). Control of GHG emission at the microbial community level. *Waste Management, 28*(4), 699–706.

Jimenez-Hornero, J. E., Santos-Duenas, I. M., & Garcia-Garcia, I. (2009). Optimization of biotechnological processes. The acetic acid fermentation. Part II: Practical identifiability analysis and parameter estimation. *Biochemical Engineering Journal, 45*(1), 7–21.

Jin, C., Wang, C., Zeng, S., & He, M. (2009). Characteristics analysis and reduction approach in relation to energy consumption in municipal wastewater treatment plants (in Chinese). *Water and Wastewater Engineering, 35*(S1), 270–274.

Jones, R., Parker, W., Zhu, H., Houweling, D., & Murthy, S. (2009). Predicting the degradability of waste activated sludge. *Water Environment Research, 81*(8), 765–771.

Juang, C. P., Whang, L. M., & Cheng, H. H. (2011). Evaluation of bioenergy recovery processes treating organic residues from ethanol fermentation process. *Bioresource Technology, 102*(9), 5394–5399.

Kampschreur, M. J., Temmink, H., Kleerebezem, R., Jetten, M. S. M., & van Loosdrecht, M. C. M. (2009). Nitrous oxide emission during wastewater treatment. *Water Research, 43*(17), 4093–4103.

Kenway, S. J. (2010). *The connection between water and energy in cities*. Canada: Montreal.

Kobya, M., Bayramoglu, M., & Eyvaz, M. (2007). Techno-economical evaluation of electrocoagulation for the textile wastewater using different electrode connections. *Journal of Hazardous Materials, 148*(1–2), 311–318.

Kowalchuk, G. A., & Stephen, J. R. (2001). Ammonia-oxidizing bacteria: A model for molecular microbial ecology. *Annual Review of Microbiology, 55*, 485–529.

Langergraber, G., Rieger, L., Winkler, S., Alex, J., Wiese, J., Owerdieck, C., et al. (2004). A guideline for simulation studies of wastewater treatment plants. *Water Science and Technology, 50*(7), 131–138.

Larrea, L., Albizuri, J., Abad, A., Larrea, A., & Zalakain, G. (2007). Optimizing and modelling nitrogen removal in a new configuration of the moving-bed biofilm reactor process. *Water Science and Technology, 55*(8–9), 317–327.

Le Corre, K. S., Valsami-Jones, E., Hobbs, P., & Parsons, S. A. (2009). Phosphorus recovery from wastewater by struvite crystallization: A review. *Critical Reviews in Environmental Science and Technology, 39*(6), 433–477.

Lee, T. T., Wang, F. Y., & Newell, R. B. (2006). Advances in distributed parameter approach to the dynamics and control of activated sludge processes for wastewater treatment. *Water Research, 40*(5), 853–869.

Li, Z., Qi, R., An, W., Mino, T., Shoji, T., Verstraete, W., et al. (2015). Simulation of long-term nutrient removal in a full-scale closed-loop bioreactor for sewage treatment: An example of Bayesian inference. *Frontiers of Environmental Science & Engineering, 9*(3), 534–544.

Liu, Y. C., Shi, H. C., Shi, H. M., & Wang, Z. Q. (2010). Study on a discrete-time dynamic control model to enhance nitrogen removal with fluctuation of influent in oxidation ditches. *Water Research, 44*(18), 5150–5157.

Liwarska-Bizukojc, E., & Biernacki, R. (2010). Identification of the most sensitive parameters in the activated sludge model implemented in BioWin software. *Bioresource Technology, 101*(19), 7278–7285.

Liwarska-Bizukojc, E., Olejnik, D., Biernacki, R., & Ledakowicz, S. (2011). Calibration of a complex activated sludge model for the full-scale wastewater treatment plant. *Bioprocess and Biosystems Engineering, 34*(6), 659–670.

Logan, B. E. (2004). Extracting hydrogen electricity from renewable resources. *Environmental Science and Technology, 38*(9), 160A–167A.

Logan, B. E., Hamelers, B., Rozendal, R. A., Schrorder, U., Keller, J., Freguia, S., et al. (2006). Microbial fuel cells: Methodology and technology. *Environmental Science and Technology, 40*(17), 5181–5192.

Lundin, M., Bengtsson, M., & Molander, S. (2000). Life cycle assessment of wastewater systems: Influence of system boundaries and scale on calculated environmental loads. *Environmental Science and Technology, 34*(1), 180–186.

Makinia, J. (2010). *Mathematical modelling and computer simulation of activated sludge systems*. IWA Publishing.

Massey, M. S., Davis, J. G., Ippolito, J. A., & Sheffield, R. E. (2009). Effectiveness of recovered magnesium phosphates as fertilizers in neutral and slightly alkaline soils. *Agronomy Journal, 101*(2), 323–329.

McCarty, P. L., Bae, J., & Kim, J. (2011). Domestic wastewater treatment as a net energy producer-can this be achieved? *Environmental Science and Technology, 45*(17), 7100–7106.

McLean, K. A. P., & McAuley, K. B. (2012). Mathematical modelling of chemical processesuobtaining the best model predictions and parameter estimates using identifiability and estimability procedures. *Canadian Journal of Chemical Engineering, 90*(2), 351–366.

Metcalf and Eddy. (2003). *Wastewater engineering: Treatment and reuse*. New York: McGraw Will.

MFE. (2003). *New Zealand greenhouse gas inventory*. Wellington: Ministry for the Environment.

Middlebrooks, E. J., Middlebrooks, C. H., & Reed, S. C. (1952). Energy requirement for small wastewater treatment system. *Water Pollution Control Federation, 53*(7), 1172–1197.

Ni, B. J., & Yu, H. Q. (2010). Mathematical modeling of aerobic granular sludge: A review. *Biotechnology Advances, 28*(6), 895–909.

Ni, B. J., & Yu, H. Q. (2012). Microbial products of activated sludge in biological wastewater treatment systems: A critical review. *Critical Reviews in Environmental Science and Technology, 42*(2), 187–223.

Nielsen, P. H., Saunders, A. M., Hansen, A. A., Larsen, P., & Nielsen, J. L. (2012). Microbial communities involved in enhanced biological phosphorus removal from wastewater—A model system in environmental biotechnology. *Current Opinion in Biotechnology, 23*(3), 452–459.

Ong, S. K., Koh, T. H., & Nee, A. Y. C. (2001). Assessing the environmental impact of materials processing techniques using an analytical hierarchy process method. *Journal of Materials Processing Technology, 113*(1–3), 424–431.

Owen, W. F. (1982). *Energy in wastewater treatment*. Englewood Cliffs, NJ: Prentice hall.

Papa, M., Alfonsin, C., Moreira, M. T., & Bertanza, G. (2016). Ranking wastewater treatment trains based on their impacts and benefits on human health: A "Biological Assay and Disease" approach. *Journal of Cleaner Production, 113,* 311–317.

Pasqualino, J. C., Meneses, M., Abella, M., & Castells, F. (2009). LCA as a decision support tool for the environmental improvement of the operation of a municipal wastewater treatment plant. *Environmental Science and Technology, 43*(9), 3300–3307.

Patsios, S. I., & Karabelas, A. J. (2010). A review of modeling bioprocesses in membrane bioreactors (MBR) with emphasis on membrane fouling predictions. *Desalination and Water Treatment, 21*(1–3), 189–201.

Petzet, S., & Cornel, P. (2011). Towards a complete recycling of phosphorus in wastewater treatment—Options in Germany. *Water Science and Technology, 64*(1), 29–35.

Phillips, H. M., Sahlstedt, K. E., Frank, K., Bratby, J., Brennan, W., Rogowski, S., et al. (2009). Wastewater treatment modelling in practice: A collaborative discussion of the state of the art. *Water Science and Technology, 59*(4), 695–704.

Plosz, B. G., Liltved, H., & Ratnaweera, H. (2009). Climate change impacts on activated sludge wastewater treatment: A case study from Norway. *Water Science and Technology, 60*(2), 533–541.

Poulsen, T. G., & Hansen, J. A. (2009). Assessing the impacts of changes in treatment technology on energy and greenhouse gas balances for organic waste and wastewater treatment using historical data. *Waste Management and Research, 27*(9), 861–870.

Rabaey, K., Angenent, L., Schroder, U., & Keller, J. (2010). *Bioelectrochemical systems: From extracellular electron transfer to biotechnological application*. IWA publishing.

Reichert, P. (1994). AQUASIM—A tool for simulation and data-analysis of aquatic systems. *Water Science and Technology, 30*(2), 21–30.

Ren, N. Q., Guo, W. Q., Liu, B. F., Cao, G. L., & Ding, J. (2011). Biological hydrogen production by dark fermentation: Challenges and prospects towards scaled-up production. *Current Opinion in Biotechnology, 22*(3), 365–370.

Renou, S., Thomas, J. S., Aoustin, E., & Pons, M. N. (2008). Influence of impact assessment methods in wastewater treatment LCA. *Journal of Cleaner Production, 16*(10), 1098–1105.

Rosso, D., & Bolzonella, D. (2009). Carbon footprint of aerobic biological treatment of winery wastewater. *Water Science and Technology, 60*(5), 1185–1189.

Rosso, D., Libra, J. A., Wiehe, W., & Stenstrom, M. K. (2008). Membrane properties change in fine-pore aeration diffusers: Full-scale variations of transfer efficiency and headloss. *Water Research, 42*(10–11), 2640–2648.

Rosso, D., & Stenstrom, M. K. (2008). The carbon-sequestration potential of municipal wastewater treatment. *Chemosphere, 70*(8), 1468–1475.

Roux, P., Boutin, C., Risch, E., & Heduit, A. (2010). *Life cycle environmental assessment (LCA) of sanitation systems including sewerage: Case of vertical flow constructed wetlands versus activated sludge*. Italy: Venise.

Rullkens, W. H., Van Dijk, L., Temmink, B. G., & Man, A. (2009). *Innovative sludge treatment scenarios to optimize energy, phosphate and ammonia recovery*. China: Harbin.

Salem, S., Berends, D., Heijnen, J. J., & van Loosdrecht, M. C. M. (2002). Model-based evaluation of a new upgrading concept for N-removal. *Water Science and Technology, 45*(6), 169–176.

Santero, N. J., Masanet, E., & Horvath, A. (2011). Life-cycle assessment of pavements. Part I: Critical review. *Resources Conservation and Recycling, 55*(9–10), 801–809.

Schmidt, W. P., & Sullivan, J. (2002). Weighting in life cycle assessments in a global context. *International Journal of Life Cycle Assessment, 7*(1), 5–10.

Seviour, R. J., Mino, T., & Onuki, M. (2003). The microbiology of biological phosphorus removal in activated sludge systems. *FEMS Microbiology Reviews, 27*(1), 99–127.

Shahabadi, M. B., Yerushalmi, L., & Haghighat, F. (2009). Impact of process design on greenhouse gas (GHG) generation by wastewater treatment plants. *Water Research, 43*(10), 2679–2687.

Shahabadi, M. B., Yerushalmi, L., & Haghighat, F. (2010). Estimation of greenhouse gas generation in wastewater treatment plants—Model development and application. *Chemosphere, 78*(9), 1085–1092.

Shanahan, J. W., & Semmens, M. I. (2004). Multipopulation model of membrane-aerated biofilms. *Environmental Science and Technology, 38*(11), 3176–3183.

Shizas, I., & Bagley, D. M. (2004). Experimental determination of energy content of unknown organics in municipal wastewater streams. *Journal of Energy Engineering-Asce, 130*(2), 45–53.

Show, K. Y., Lee, D. J., & Chang, J. S. (2011). Bioreactor and process design for biohydrogen production. *Bioresource Technology, 102*(18), 8524–8533.

Simsek, H., Kasi, M., Wadhawan, T., Bye, C., Blonigen, M., & Khan, E. (2012). Fate of dissolved organic nitrogen in two stage trickling filter process. *Water Research, 46*(16), 5115–5126.

Smith, B. R. (2009). Re-thinking wastewater landscapes: Combining innovative strategies to address tomorrow's urban wastewater treatment challenges. *Water Science and Technology, 60*(6), 1465–1473.

Soares, S. R., Toffoletto, L., & Deschenes, L. (2006). Development of weighting factors in the context of LCIA. *Journal of Cleaner Production, 14*(6–7), 649–660.

Sun, M., Sheng, G. P., Zhang, L., Xia, C. R., Mu, Z. X., Liu, X. W., et al. (2008). An MEC-MR-coupled system for biohydrogen production from acetate. *Environmental Science and Technology, 42*(21), 8095–8100.

Sweeney, M. W., & Kabouris, J. C. (2004). Modeling, instrumentation, automation, and optimization of wastewater treatment facilities. *Water Environment Research, 76*(6), 1432–1478.

Tangsubkul, N., Beavis, P., Moore, S. J., Lundie, S., & Waite, T. D. (2005). Life cycle assessment of water recycling technology. *Water Resources Management, 19*(5), 521–537.

Thomsen, M., & Lyck, E. (2005). *Emission of CH_4 and N_2O from wastewater treatment plants (6B)*. Denmark: National Environmental Research Institute, Ministry of the Environment.

Thygesen, A. M., Wernberg, O., Skou, E., & Sommer, S. G. (2011). Effect of incineration temperature on phosphorus availability in bio-ash from manure. *Environmental Technology, 32*(6), 633–638.

Verstraete, W., & Vlaeminck, S. E. (2011). ZeroWasteWater: Short-cycling of wastewater resources for sustainable cities of the future. *International Journal of Sustainable Development and World Ecology, 18*(3), 253–264.

Wang, X., Feng, Y. J., Liu, J., Lee, H., Li, C., Li, N., et al. (2010). Sequestration of CO_2 discharged from anode by algal cathode in microbial carbon capture cells (MCCs). *Biosensors & Bioelectronics, 25*(12), 2639–2643.

Wesner, G. M., & Culp, G. L. (1978). *Energy conservation in municipal wastewater treatment, U.S.* Washington, DC: EPA.

Wong, B. T., Show, K. Y., Lee, D. J., & Lai, J. Y. (2009). Carbon balance of anaerobic granulation process: Carbon credit. *Bioresource Technology, 100*(5), 1734–1739.

Xin, C. H., Addy, M. M., Zhao, J. Y., Cheng, Y. L., Cheng, S. B., Mu, D. Y., et al. (2016). Comprehensive techno-economic analysis of wastewater-based algal biofuel production: A case study. *Bioresource Technology, 211,* 584–593.

Yang, S. (1984). Energy consumption relative to municipal wastewater treatment plants (in Chinese). *Water and Wastewater Engineering, 10*(6), 15–19.

Yang, Q., Gu, S. B., Peng, Y. Z., Wang, S. Y., & Liu, X. H. (2010). Progress in the development of control strategies for the SBR process. *Clean-Soil Air Water, 38*(8), 732–749.

Yang, Y., Yang, J. K., Zuo, J. L., Li, Y., He, S., Yang, X., et al. (2011). Study on two operating conditions of a full-scale oxidation ditch for optimization of energy consumption and effluent quality by using CFD model. *Water Research, 45*(11), 3439–3452.

Yang, L., Zeng, S., Ju, Y., He, M., & Chen, J. (2008). Statistics analysis and quantitative identification of energy consumption patterns relative to municipal wastewater treatment plants in China (in Chinese). *Water and Wastewater Engineering, 34*(10), 42–45.

Yuan, Z. G., Pratt, S., & Batstone, D. J. (2012). Phosphorus recovery from wastewater through microbial processes. *Current Opinion in Biotechnology, 23*(6), 878–883.

Zhang, X., & Wang, X. (2009). Environmental benefits analysis of municipal wastewater treatment (in Chinese). *Chinese Journal of Environmental Engineering, 3*(5), 861–863.

Zhang, Z., & Wilson, F. (2000). Life-cycle assessment of a sewage-treatment plant in South-East Asia. *Journal of the Chartered Institution of Water and Environmental Management, 14*(1), 51–56.

Chapter 3
Life Cycle Inventory Analysis of Typical Wastewater Treatment Chains

3.1 Overview

It is widely known that the purpose of constructing and commissioning municipal wastewater treatment plants (WWTPs) is to reduce the pollution and eutrophication of receiving waters caused by wastewater discharge. The recent rapid growth of the world population and increasingly stringent demand for water quality have become new driving forces in the research and development of new municipal wastewater treatment processes and the upgrade and reconstruction of existing processes. However, an increase in the wastewater treatment efficiency will inevitably lead to an accelerated depletion of resources and may even result in other environmental problems such as pollution transfer. Furthermore, the additive effects of the aforementioned problems may create more severe environmental burdens. Nevertheless, the quality of the water environment in local regions remains the topic of greatest concern in the water industry; however, other environmental burdens have not received the attention they deserve. Furthermore, different wastewater treatment processes may cause different environmental effects. However, research on wastewater treatment technologies and other processes in China has long focused on the process performance and there is a lack of analyses of the environmental impacts of different wastewater treatment systems that are systematic and comparable.

The life cycle assessment (LCA) method is a systematic and quantitative assessment technique that puts emphasis on the environmental impact of a product or service throughout its entire life cycle, that is, "from cradle to grave." When the LCA concept is used in the systematic analysis of typical municipal wastewater treatment processes to identify the types of environmental burden and key links in the operation of different process combinations, an in-depth understanding of the environmental impacts of different wastewater treatment processes can be gained and the results can serve as references for the investment in and construction of new WWTPs and the upgrade and reconstruction of existing plants. In addition, the use of

X. Wang, *Energy Consumption, Chemical Use and Carbon Footprints of Wastewater Treatment Alternatives*, Springer Theses,
https://doi.org/10.1007/978-981-13-5983-5_3

the LCA approach can also provide ideas for the design of novel assessment systems for wastewater treatment in subsequent chapters.

However, the application of theory and knowledge of multiple disciplines, such as chemistry, physics, mathematics, toxicology, statistics, and environmental studies, is required for the LCA method to focus on the environmental burdens of the studied subject with respect to natural resources, nonliving components of ecosystems, and human health and ecotoxicity. Researchers must have an in-depth understanding of the entire life cycle of the study subject and spend a considerable amount of time on certain aspects such as system boundary delineation, model selection, and data mining. The objectives of this chapter are to analyze the environmental burdens of typical wastewater treatment processes and identify the sources and main contributing units of environmental impact of different process combinations. The LCA method was used as guidance for this study. However, the conventional LCA method was not applied; instead, only an environmental impact inventory analysis was performed. This chapter focuses on the accurate and rapid analysis of the sources and key links contributing to environmental burdens in typical wastewater treatment processes. This approach is the main focus of this chapter.

3.2 Methods for the Study of Environmental Impacts of Wastewater Treatment Processes

3.2.1 Study Subjects

The typical municipal wastewater treatment process is as follows. First, wastewater is transported via lifting pump stations into a WWTP. After pretreatments, such as screening and sedimentation, the wastewater enters a primary clarifier for primary processing. The effluent obtained after clarification flows into a bioreactor for secondary biological treatment. If there are special requirements, the effluent from the secondary clarifier can be utilized as a resource after undergoing advanced treatment or it can be directly discharged into receiving waters. After thickening (or digestion) and dewatering, the sludge in primary and secondary clarifiers undergoes further processing such as composting, drying, or incineration. The treatment process in most of the existing WWTPs consists of pretreatment, primary treatment, secondary treatment, and sludge thickening and dewatering. Secondary biological treatment processes are the core of municipal WWTPs. Its main function is to remove colloidal and dissolved organic matter and major nutrients (nitrogen and phosphorus) from the wastewater. Therefore, this chapter mainly focuses on secondary biological treatment processes in municipal WWTPs. Current systems used for secondary biological treatment consist of different spatial or temporal combinations of anoxic, anaerobic, and aerobic treatment units (Kang et al. 2011; Mulkerrins et al. 2000; Wang et al. 2011; Zeng et al. 2010; Zhou et al. 2011). The most representative units are the anaerobic/anoxic/oxic (A^2/O) process (spatial arrangement) and sequencing batch reactor (SBR) process (temporal arrangement). In this chapter, the spatial arrangements were investigated by studying six combinations of anaerobic, anoxic, and aerobic biological treatment processes that are commonly used in WWTPs (see Sect. 3.3 for details).

3.2.2 System Boundaries and Model Construction Concepts

In the present work, the basic concepts of LCA were used to determine system boundaries. Wastewater treatment processes are discussed in this chapter. Wastewater collection and transportation processes are not focused on because the wastewater collection via pipelines does not influence the secondary biological processes in WWTPs. Therefore, the influent flow to a municipal WWTP and the effluent flow from the secondary clarifier and the dewatering and transportation of sludge were set as the start and end points of the study system, respectively. The system includes the main treatment segments of water and sludge in the municipal WWTP. The water segment includes the anaerobic, aerobic, and anoxic biological treatment units and secondary clarifier. A primary clarifier was not used in the system to ensure sufficient carbon sources for subsequent biological processes. The sludge segment mainly includes the thickening, anaerobic digestion, and dewatering of excess sludge, while the dosing of an external carbon source and alkaline solution are simple nodes in the system.

The life cycle boundaries are usually specified when the LCA method is used for system analysis. A complete WWTP life cycle includes the entire process starting from the acquisition of initial resources, energy sources, and raw materials from nature by extraction, processing, construction, and long-term operation and ending with scrapping and demolition after the expiration of the operation period of the WWTP. In short, the entire "cradle to grave" life cycle of WWTPs can be divided into three stages: (1) material production and construction; (2) operation; and (3) scrapping and demolition. The best methods based on which the environmental impact of a WWTP can be understood are the quantification and interpretation of environmental impacts during different stages of the entire life cycle of the WWTP. However, Lundie et al. (2004) found that the environmental impact of the operational stage is larger than that of the construction and demolition stages. Tillman et al. (1998) also reported that the environmental burdens of different wastewater treatment schemes during the construction stage are comparable, while they observed differences of environmental burdens during the operational stage of the wastewater treatment systems. From another perspective, multiple material and energy flows are accounted for during the construction and demolition stages of municipal WWTPs. Therefore, the definition and quantification of these two life cycle stages will undoubtedly increase the complexity of the study system and uncertainty of the results.

In view of the above-mentioned analysis, the environmental impacts of the construction and demolition stages of wastewater treatment systems were not considered in this study to accentuate the focus of the study, simplify the research system, and reduce the uncertainty of the results; that is, only the environmental burdens based on wastewater processing were quantified and analyzed.

3.2.3 Function and Functional Unit

The functional unit is a measure of the function of the product system, and its basic role is to provide a mathematical standard for the standardized measurement of the inputs and outputs during LCA. The functional unit of a WWTP is the treatment scale, that is, the treatment capacity of the WWTP. When LCA is used to assess different wastewater treatment processes, the functional unit must be determined first. The main purpose of defining the functional unit is to standardize the input and output streams of the wastewater treatment systems. This provides a reference standard for relevant input and output stream data to ensure the comparability of the assessment results. The wastewater inflow rate of the system studied in this chapter is 200,000 tons/day, and the functional unit used for the assessment is the environmental impact based on the treatment of 1 ton of wastewater per day.

3.2.4 Methods for the Analysis of the Environmental Impact

For a long time, the main factors considered for the construction of municipal WWTPs have been the processing functions, while the resultant problems, such as resource consumption, energy consumption, and pollutant transfer, have usually been neglected. Recently, researchers reported that the chemical consumption, energy consumption, and greenhouse gas (GHG) emissions of wastewater treatment in municipal WWTPs are closely linked to adverse environmental impacts such as global warming, ozone depletion, acidification, and ecotoxicity (Foley et al. 2010a; Horne et al. 2009; Shaw et al. 2011). Therefore, the chemical consumption, energy consumption, and GHG emissions were selected as macroscopic representations of the environmental impact of WWTPs in this chapter and quantitative inventory analysis of the resource and energy use and GHG discharged to the environment was performed. For the inventory analysis, which was performed in accordance with LCA guidelines (Guinee 2002), the inventory was prepared through data collection and confirmation, linking of data and unit processes, linking of data and the functional unit, data merging, system boundary optimization, and data feedback.

3.3 Wastewater Treatment Process Simulation and Data Validation

3.3.1 *Process Flows*

The processing capacity of the WWTP, chemical oxygen demand (COD) of the influent, total nitrogen (TN), and total phosphorus (TP) was set to 200,000 m^3/d, 500 mg/L, 50 mg/L, and 12 mg/L, respectively. After entering the WWTP, the wastewater reaches the secondary biological process unit and subsequently enters the secondary clarifier. When the effluent of the secondary clarifier complies with the relevant preset discharge standard, it is discharged to the receiving waters. Concurrently, excess sludge discharged from the secondary clarifier is transported out of the WWTP after thickening, digestion, and dewatering. The detailed process is shown in Fig. 3.1. In addition, to investigate the impact of different wastewater discharge standards on the environmental burdens caused by the WWTP, the Class 2, Class 1B, and Class 1A standards of the Pollutant Discharge Standard of Municipal Wastewater Treatment Plants (GB18918-2002) were implemented as the treatment standards. Based on these standards, the, respectively, effluent quality requirements are as follows: Class 2: COD $\leq$ 100 mg/L, NH_3–N $\leq$ 25 mg/L, TP $\leq$ 3 mg/L; Class 1B: COD $\leq$ 60 mg/L, NH_3–N $\leq$ 8 mg/L, TN $\leq$ 20 mg/L, TP $\leq$ 1.5 mg/L; and Class 1A: COD $\leq$ 50 mg/L, NH_3–N $\leq$ 5 mg/L, TN $\leq$ 15 mg/L, TP $\leq$ 1 mg/L.

3.3.2 *Alternative Wastewater Treatment Scenarios*

Based on the given influent and effluent quality requirements and wastewater treatment process design approaches (Metcalf and Eddy 2003), six typical alterna-

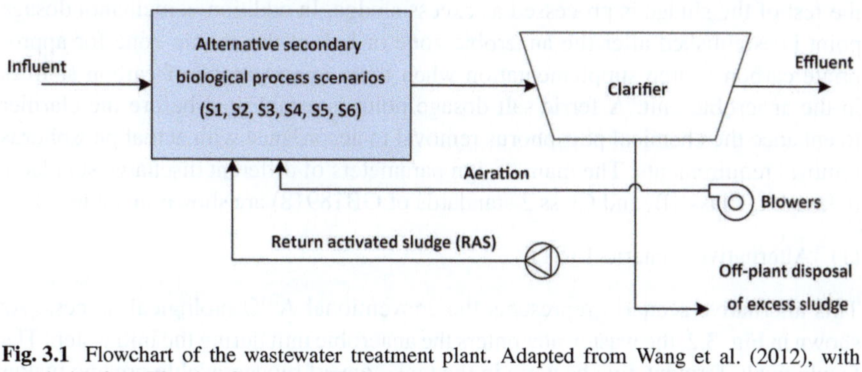

Fig. 3.1 Flowchart of the wastewater treatment plant. Adapted from Wang et al. (2012), with permission from Elsevier, Copyright (2012). *Note* S1–S6 are the numbers for the different alternative treatment scenarios (see Sect. 3.3.2 for details)

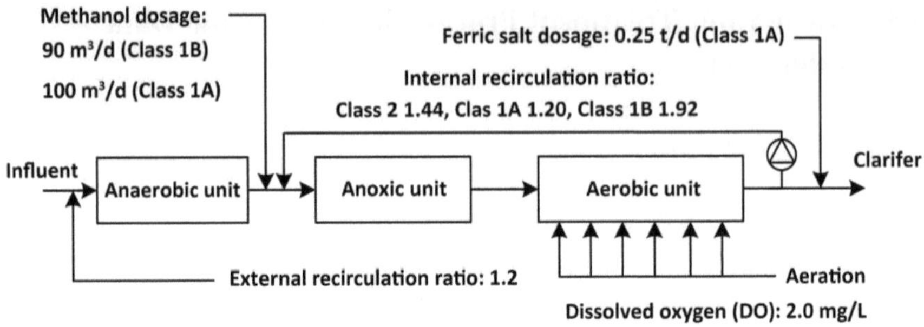

Fig. 3.2 Flow scheme of alternative scenario 1 (S1). Adapted from Wang et al. (2012), with permission from American Chemical Society, Copyright (2012)

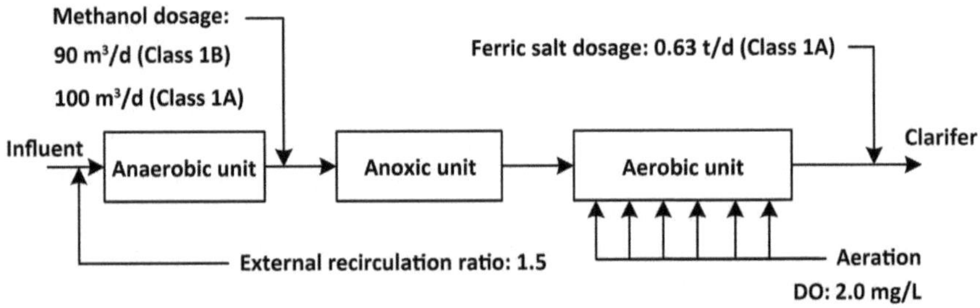

Fig. 3.3 Flow scheme of alternative scenario 2 (S2). Adapted from Wang et al. (2012), with permission from American Chemical Society, Copyright (2012)

tive wastewater treatment scenarios including different combinations of anaerobic, anoxic, and aerobic units (Figs. 3.2, 3.3, 3.4, 3.5, 3.6, and 3.7) were designed. Each alternative scenario comprises an anaerobic tank, anoxic tank, aerobic tank, and secondary clarifier connected in series. A portion of the activated sludge discharged from the secondary clarifier is returned to the anaerobic tank or anoxic tank, while the rest of the sludge is processed as excess sludge. In addition, a methanol dosage point is established after the anaerobic zone or before the anoxic zone for appropriate carbon source supplementation when there is a shortage of carbon sources in the anaerobic unit. A ferric salt dosage point is established before the clarifier to enhance the chemical phosphorus removal in accordance with actual phosphorus removal requirements. The main design parameters of different discharge standards (Class 1A, Class 1B, and Class 2 standards of GB18918) are shown in Table 3.1.

(1) Alternative scenario 1 (S1)

This alternative scenario represents the conventional A^2/O biological process. As shown in Fig. 3.2, the wastewater enters the anaerobic unit during the initial step. The facultatively fermentative bacteria in the tank convert biodegradable organic matter to fermentation products such as volatile organic acids; polyphosphate-accumulating organisms (PAOs) break down the polyphosphate salts accumulated within their cells,

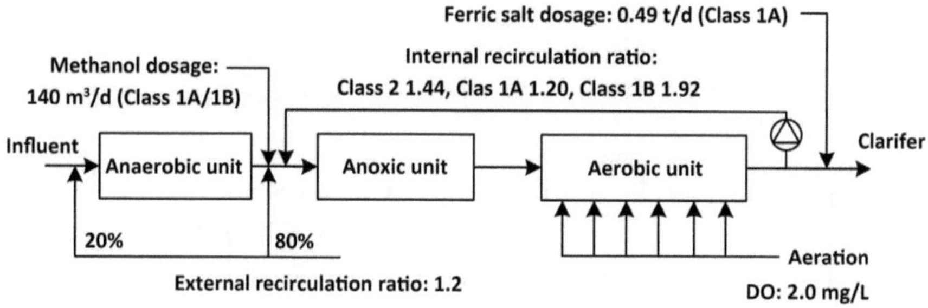

Fig. 3.4 Flow scheme of alternative scenario 3 (S3). Adapted from Wang et al. (2012), with permission from American Chemical Society, Copyright (2012)

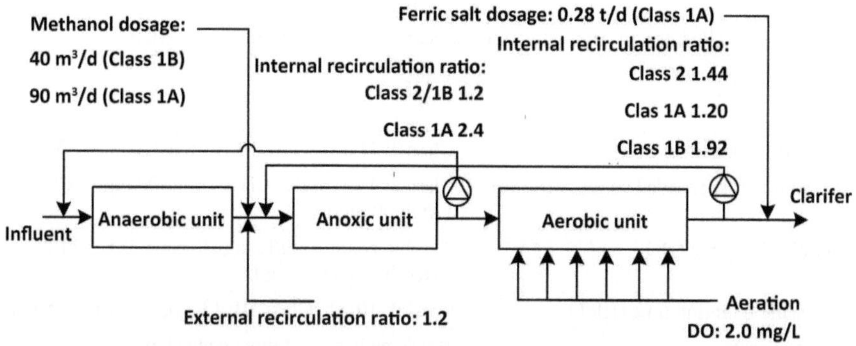

Fig. 3.5 Flow scheme of alternative scenario 4 (S4). Adapted from Wang et al. (2012), with permission from American Chemical Society, Copyright (2012)

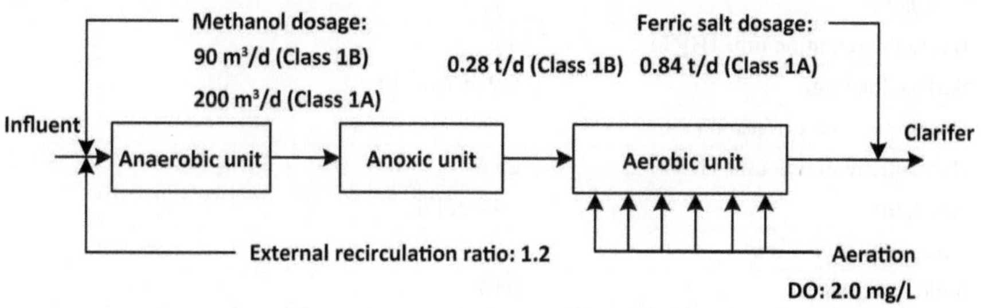

Fig. 3.6 Flow scheme of alternative scenario 5 (S5). Adapted from Wang et al. (2012), with permission from American Chemical Society, Copyright (2012)

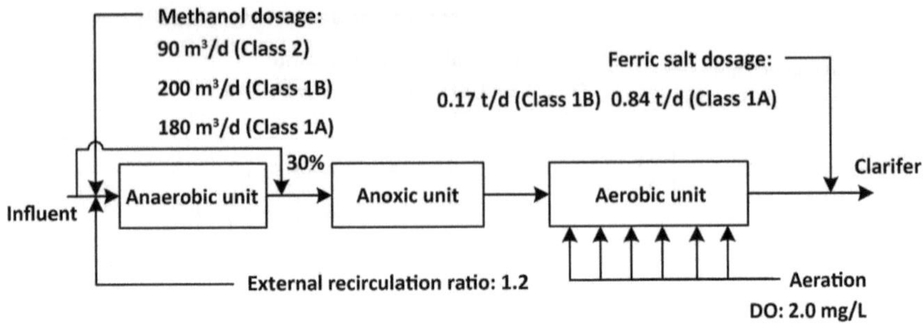

Fig. 3.7 Flow scheme of alternative scenario 6 (S6). Adapted from Wang et al. (2012), with permission from American Chemical Society, Copyright (2012)

Table 3.1 Detailed design parameters for alternative wastewater treatment scenarios

Design parameter	Value
Biological unit	
Solid residence time (SRT)	Class 2: 11.1 d, Class 1B: 13.8 d, Class A: 15.3 d
Mixed liquor suspended solids (MLSSs)	Class 2: 4000 mg/L, Class 1B: 4500 mg/L, Class 1A: 5000 mg/L
Dissolved oxygen (DO)	Aerobic zone: 2 mg/L, anoxic zone: 0.5 mg/L, anaerobic zone: 0 mg/L
Hydraulic retention time (HRT)	Class 2: 10.7 h, Class 1B: 13.7 h, Class 1A: 15.6 h
External carbon source	100% methanol solution, equivalent to 1.188 kg COD/L
Ferric salts for phosphorus removal	Flow rate: 2% of the influent flow rate
Clarifier	
Hydraulic retention time (HRT)	4 h
Surface loading	0.9 $m^3/(m^2\ h)$
Anaerobic sludge digestion unit	
Hydraulic retention time (HRT)	25 d
Alkalinity	30 kmol/d
Sludge-thickening unit	
Solid capture	95%
Water content after thickening	94%
Sludge dewatering unit	
Solid capture	95%
Water content after dewatering	85%

Adapted from Wang et al. (2012), with permission from Elsevier, Copyright (2012)

and the released energy facilitates the survival of PAOs under anaerobic conditions and enables PAOs to absorb organic compounds with low molecular weight in the environment and to store the compounds as polymers within their cells. The effluent from the anaerobic unit enters the anoxic unit, where denitrifying bacteria make use of nitrates in the mixed liquor recirculation (MLR) stream and biodegradable organic matter in wastewater to perform denitrification, thereby achieving simultaneous COD removal and denitrification. Subsequently, the effluent from the anoxic unit enters the aerobic unit, where PAOs concurrently utilize the residual biodegradable organic matter in wastewater and degrade the polymers accumulated within their cells to produce energy for their growth and reproduction. In addition, the PAOs also absorb phosphates dissolved in the surrounding environment and store the phosphates as polyphosphates within their cells. Because the organic carbon in the influent to the wastewater treatment process is utilized by the PAOs and denitrifying bacteria in the anaerobic and anoxic units, respectively, the concentration of organic carbon in the influent to the aerobic unit is extremely low. This favors the growth of autotrophic nitrifying bacteria, which convert ammonia nitrogen to nitrates through nitrification.

(2) Alternative scenario 2 (S2)

In this alternative scenario, which is a modification of the conventional A^2/O process, MLR is eliminated and the return activated sludge (RAS) ratio is increased based on S1 (Fig. 3.3). Firstly, an anoxic zone naturally forms when the RAS is mixed with the influent wastewater. The denitrifying bacteria in sludge utilize the organic matter in the influent as a carbon source for denitrification, thereby leading to the rapid consumption of nitrates in the RAS and formation of a strictly anoxic environment in the subsequent zone. The denitrification effects of this alternative scenario include the endogenous denitrification of microbes in the aeration tank and denitrification by the RAS in the anaerobic zone based on the utilization of organic matter in the influent wastewater as a carbon source.

(3) Alternative scenario 3 (S3)

In this alternative scenario, which is another modification of the conventional A^2/O process, the distribution of RAS has been changed. As shown in Fig. 3.4, 20% of the RAS is returned to the anaerobic unit, while the remaining 80% of the RAS are returned to the anoxic unit, which reduces the impact of nitrates on the phosphorus release by PAOs. Therefore, only a small portion of RAS undergoes complete anaerobic and aerobic biological processes, while most of the RAS directly enters the anoxic and aerobic environments without going through the anaerobic unit.

(4) Alternative scenario 4 (S4)

The University of Cape Town proposed a scenario similar to that shown in Fig. 3.5, which is known as the University of Cape Town (UCT) process. This scenario differs from the conventional A^2/O process in that the RAS from the final clarifier is not returned to the anaerobic unit of the A^2/O process; instead, it is returned to the anoxic unit. Consequently, the adverse impact on the phosphorus removal efficiency of the system induced by the entry of nitrate-nitrogen into the anaerobic tank and subsequent disturbance of the anaerobic conditions of the tank can be avoided. At the same time, MLR from the anoxic unit to the anaerobic unit is introduced in this scenario. Because the denitrification effects in the anoxic unit lead to a substantial reduction in the nitrate-nitrogen concentration, the MLR from the anoxic unit will not disturb the anaerobic conditions of the anaerobic tank. Based on this process, the nitrate load of the anaerobic unit reduces; however, the amount of recirculation increases.

(5) Alternative scenario 5 (S5)

In this process, the positions of the anaerobic and anoxic units of the conventional A^2/O process are switched; that is, the anoxic unit is placed before the anaerobic unit, resulting in a reversed A^2/O process (Fig. 3.6). The characteristics of this process are as follows: After anaerobic phosphorus release, PAOs directly enter the aerobic environment, which leads to better utilization of the phosphorus uptake of kinetics that formed under anaerobic conditions, that is, a "starvation effect." The whole RAS undergoes a complete phosphorus release and uptake process, which creates a "population effect." The anoxic zone is located in the first unit of the process; the denitrification process therefore gains priority during carbon source acquisition, which further enhances the denitrification capacity of the system. By adopting appropriate engineering measures, the RAS and MLR flows can be combined in a recirculation system to simplify the process.

(6) Alternative scenario 6 (S6)

In this process, which is shown in Fig. 3.7, the influent enters the treatment process at multiple points based on the reversed A^2/O process. A portion of the wastewater enters the anoxic unit, while the rest of the wastewater enters the anaerobic unit, thereby providing separate carbon sources for denitrification and biological phosphorus removal. In addition, when the RAS and a portion of the wastewater enter the anoxic unit, the nitrates in the RAS undergo denitrification, thus ensuring completely anaerobic conditions in the subsequent anaerobic unit.

3.3.3 *Modeling Process*

To achieve better WWTP simulations and wastewater and sludge treatment processes and acquire complete basic data to support inventory analysis, the research work in this chapter has been focused on the numerical modeling and simulation of an entire WWTP. The models involved in the modeling of the whole WWTP process mainly include the activated sludge model, anaerobic digestion model, chemical clarification model, and physical settling model (Makinia 2010).

3.3.3.1 Activated Sludge Model (ASM)

The mathematical descriptions of the wastewater treatment processes in this chapter are primarily based on the framework of the Barker–Dold model (Barker and Dold 1997b). Four types of activated microbes, that is, ordinary heterotrophs (Z_{bh}), ammonia-oxidizing bacteria (Z_{aob}), nitrite-oxidizing bacteria (Z_{nob}), and polyphosphate-accumulating organisms (Z_{bp}), and associated metabolic processes have been considered in the model.

A total of 11 biochemical reactions describe the growth and decay of Z_{bh} under aerobic and anoxic conditions. Its growth substrates include readily biodegradable organic carbon (S_{bsc}), acetic acid (S_{bsa}), and propionic acid (S_{bsp}). The Z_{aob} gains energy from the oxidation of ammonia to nitrites, which is used for the synthesis of organic materials from inorganic carbon (carbon dioxide), with ammonia being the nitrogen source. The metabolism of Z_{aob} can be described by two biochemical reactions. The energy for Z_{nob} is gained through the oxidation of nitrites to nitrates, which is used for the synthesis of organic materials from inorganic carbon, with ammonia being the nitrogen source for synthesis. The growth and decay of Z_{nob} can be described by two biochemical reactions. A total of 15 biochemical kinetic processes describe the growth and decay of Z_{bp} under all conditions including aerobic and anoxic growth, volatile fatty acid (S_{bsa} or S_{bsp}) storage, polymer (S_{pha}) formation, and polyphosphate (*PP-lo*) degradation. In addition, the death–regeneration model proposed by Dold et al. (1980) was used in the present work to describe the decay and formation of endogenous respiration products of Z_{bh}, Z_{aob}, and Z_{nob}, while the endogenous respiration model proposed by McKinney(1960) was used to describe the decay of Z_{bp}.

In addition, carbon sources in wastewater are usually inadequate because of the extremely stringent requirements on TN. Therefore, methanol dosage is carried out on carbon sources in the anoxic zone to facilitate the denitrification process. Consequently, in the conceptual model, a type of bacteria that specifically utilizes methanol (Z_{bmeth}) was separated from other bacteria in Z_{bh} due to different growth characteristics (growth rate and unit yield coefficient). A total of three kinetic processes describe the growth and decay of S_{bmeth} under anoxic conditions. At the same time, it is assumed that S_{bmeth} only grows and converts nitrites or nitrates to nitrogen by denitrification under anoxic conditions and ammonia is used as the nitrogen source for the growth of S_{bmeth}.

In the Barker–Dold model, a total of seven biochemical kinetic processes describe the conversion of organic components and nitrogen and phosphorus components under microbial action including the (1) hydrolysis of slowly biodegradable particulate organic matter (X_{sp}) to readily biodegradable organic matter (S_{bsc}); (2) hydrolysis of slowly biodegradable particulate organic nitrogen (X_{on}) and organic phosphorus (X_{op}); (3) adsorption and flocculation of slowly biodegradable colloidal organic matter (X_{sc}) onto slowly biodegradable particulate organic matter (X_{sp}); (4) ammonification of soluble, readily biodegradable organic nitrogen (N_{os}); and (5) simultaneous nitrification and denitrification (at low ammonia concentrations , simultaneous nitrifi-

Table 3.2 Stoichiometric parameters used in the activated sludge model (20 °C)

Parameter	Unit	Value	Source
$Y_{\text{Zbh,Sbsc}}$	mg COD/mg COD	0.67	Barker and Dold (1997a, b)
$Y_{\text{Zbh,Sbsa}}$	mg COD/mg COD	0.40	Barker and Dold (1997a, b)
$Y_{\text{Zbh,Sbsp}}$	mg COD/mg COD	0.50	Barker and Dold (1997a, b)
Y_{Zaob}	mg COD/mg COD	0.15	Barker and Dold (1997a, b)
Y_{Znob}	mg COD/mg COD	0.09	Barker and Dold (1997a, b)
Y_{Zbp}	mg COD/mg COD	0.64	Barker and Dold (1997a, b)
Y_{Zbmeth}	mg COD/mg COD	0.40	(Makinia 2010)
$Y_{\text{P,PHA}}$	mg P/mg COD	0.95	Barker and Dold (1997a, b)
$Y_{\text{PHA,Sbsa (Sbsp)}}$	mg COD/mg COD	0.89	Barker and Dold (1997a, b)

Table 3.3 Kinetic parameters used in the activated sludge model (20 °C)

Parameter	Unit	Value		Data source
μ_{Zbh}	mg COD/mg COD d^{-1}	3.20		Barker and Dold (1997a, b)
μ_{Zaob}	mg COD/mg COD d^{-1}	0.90		Barker and Dold (1997a, b)
μ_{Znob}	mg COD/mg COD d^{-1}	0.70		Barker and Dold (1997a, b)
μ_{Zbp}	mg COD/mg COD d^{-1}	0.95		Barker and Dold (1997a, b)
$\mu_{\text{Zbp,limited}}$	mg COD/mg COD d^{-1}	0.42		Barker and Dold (1997a, b)
μ_{Zbmeth}	mg COD/mg COD d^{-1}	1.30		(Makinia 2010)
$K_{\text{s,Zbh}}$	mg COD/L	5.00		Barker and Dold (1997a, b)
$K_{\text{s,Zaob}}$	mg N/L	0.70		Barker and Dold (1997a, b)
$K_{\text{s,Znob}}$	mg N/L	0.05		Barker and Dold (1997a, b)
$K_{\text{s,Zbp}}$	mg COD/L	0.10		Barker and Dold (1997a, b)
$K_{\text{s,Zbp,limited}}$	mg COD/L	0.05		Barker and Dold (1997a, b)
$K_{\text{s,Zbmeth}}$	mg COD/L	0.50		Makinia (2010)
K_{hydro}	d^{-1}	2.10		Barker and Dold (1997a, b)
$K_{\text{d,Zbh}}$	d^{-1}	0.62[a]	0.30[b]	Barker and Dold (1997a, b)
$K_{\text{d,Zaob}}$	d^{-1}	0.17[a]	0.08[b]	Barker and Dold (1997a, b)
$K_{\text{d,Znob}}$	d^{-1}	0.17[a]	0.08[b]	Barker and Dold (1997a, b)
$K_{\text{d,Zbp}}$	d^{-1}	0.10[a]	0.04[b]	Barker and Dold (1997a, b)
$K_{\text{d,Zmeth}}$	d^{-1}	0.04[a]	0.03[b]	Makinia (2010)

[a]Values for aerobic condition
[b]Values for anoxic or anaerobic condition

cation and denitrification convert nitrites and nitrates to ammonia to promote bacteria growth). The main stoichiometric and kinetic parameters used in the activated sludge model (ASM) are shown in Tables 3.2 and 3.3, respectively.

3.3.3.2 Anaerobic Digestion Model

The anaerobic digestion model (ADM) proposed by the International Water Association is a comprehensive model for the anaerobic digestion process. However, many limitations exist when the ADM is applied to the modeling of the entire WWTP process: (1) The ADM differs from other ASMs in that it requires a transformer to support the transformations between the ASM and the ADM and the construction of such a transformer based on the conservation of mass is extremely difficult; (2) in the ADM, phosphorus is not a state variable, while phosphorus release during the anaerobic digestion process is especially important in the modeling of denitrification and phosphorus removal of wastewater; and (3) the ADM does not include nitrogen release from dead organisms, and therefore, nitrogen mass balance cannot be performed. Nitrogen release during anaerobic digestion and the return of the anaerobic digester supernatant to the raw wastewater must be specifically considered during the modeling of the entire WWTP.

Based on the considerations described above and the integration of typical ADMs (Bagley and Brodkorb 1999; Costello et al. 1991; Masse and Droste 2000; Mosey 1983), an ensemble model has been introduced in this chapter to model the anaerobic digestion process in the WWTP. This model is constructed using four types of microbe populations, that is, heterotrophic bacteria (Z_{bh}), propionic acid-utilizing bacteria (propionic acetogens; Z_{bpa}), acetoclastic methanogens (Z_{bam}), and hydrogenotrophic methanogens (Z_{bhm}), and respective metabolic behaviors. The conceptual scheme of the model is shown in Fig. 3.8. Under the action of Z_{bh}, readily biodegradable organic matter (S_{bsc}) will undergo fermentation during the anaerobic digestion process to form acetic acid (S_{bsa}), propionic acid (S_{bsp}), and H_2. The description of involved kinetic processes is consistent with that of the fermentation reaction in the anaerobic activated sludge zone. The Z_{bpa} converts S_{bsp} to S_{bsa}, CO_2, and H_2. Methane (CH_4) is produced from the combined action of Z_{bam} and Z_{bhm}, with Z_{bam} utilizing S_{bsa} for growth and Z_{bhm} using H_2 as the growth substrate.

The three metabolic microbial behaviors, that is, microbial decay, hydrolysis of particulate matter, and ammonification of organic nitrogen, are identical with the ASM described in Sect. 3.3.3.1. However, the two models are different in that anaerobic microbial decay is not significant in the ASM and the anaerobic hydrolysis rate of particulate matter is lower in the ADM compared with the ASM. Table 3.4 shows the parameters relevant to the ADM.

3.3.3.3 Model Calculations

In view of the requirements of the ensemble model described above, the BioWin simulator was employed in the present work for the modeling and simulation of the whole WWTP process. This simulation platform has been widely applied in modeling calculations and simulations in the wastewater treatment field. In addition, a series of relevant studies (Eldyasti et al. 2012; Hafez et al. 2010; Jones et al. 2009; Liu et al. 2011; Liwarska-Bizukojc and Biernacki 2010; Liwarska-Bizukojc et al.

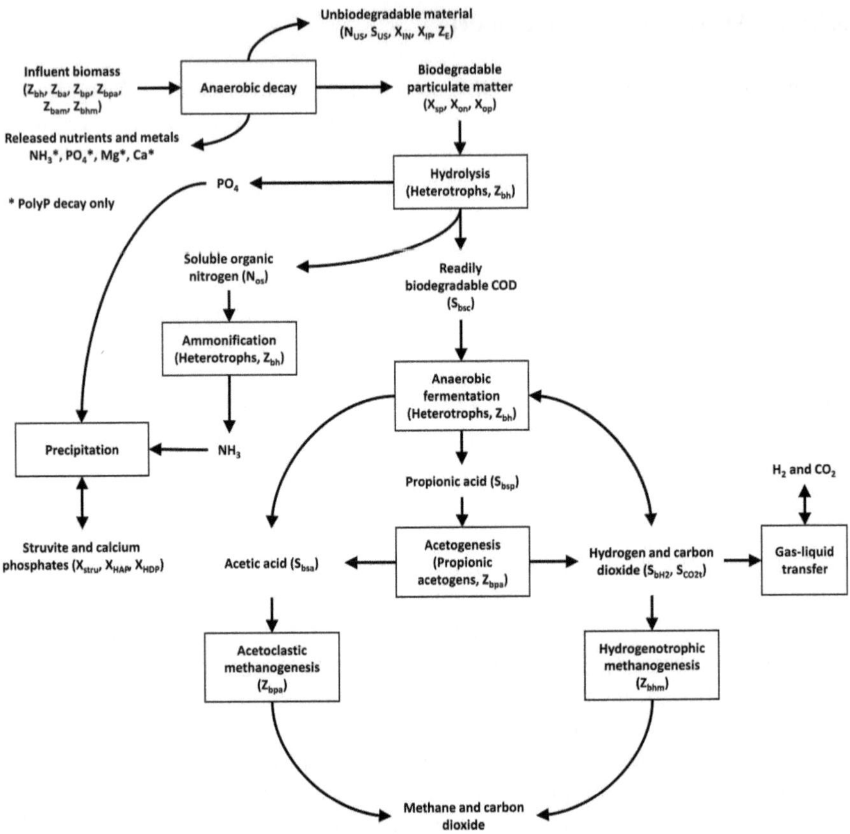

Fig. 3.8 Conceptual scheme for the anaerobic digestion model

2011; Pasztor et al. 2009) indicated a good compatibility between the ASM and the ADM of the BioWin simulator. The continuity and mass balances of organic matter, nitrogen, phosphorus, and other components can be automatically achieved, thereby eliminating the need to construct an interface model between the ASM and the ADM. The influent conditions and process design parameters mentioned in Sects. 3.3.1 and 3.3.2 were used as initial boundary values in the BioWin simulation environment, while the kinetic and stoichiometric parameters listed in Tables 3.2, 3.3, and 3.4 were also substituted into the model. In addition, the chemical clarification and physical sedimentation processes involved in the modeling of the whole WWTP process are described using modules included in BioWin. Due to space constraints, a detailed description of the modules is not presented here. For more details, refer to the official BioWin manual (Envirosim 2007).

Table 3.4 Kinetic and stoichiometric parameters in the anaerobic digestion model

Parameter	Unit	Value	Source
Kinetic parameters			
μ_{Zbpa}	d^{-1}	0.25	Bagley and Brodkorb (1999), Costello et al. (1991), Masse and Droste (2000), Mosey (1983)
μ_{Zbam}	d^{-1}	0.30	
μ_{Zbhm}	d^{-1}	1.40	
$K_{ferment}$	mg COD/L	5.00	
$K_{s,Zbh}$	d^{-1}	0.30	
$K_{s,Zbpa}$	d^{-1}	0.05	
$K_{s,Zbam}$	d^{-1}	0.13	
$K_{s,Zbhm}$	d^{-1}	0.13	
Stoichiometric parameters			
Y_{Zbh}	mg COD/mg COD	0.10	Bagley and Brodkorb (1999), Costello et al. (1991), Masse and Droste (2000), Mosey (1983)
Y_{Zbpa}	mg COD/mg COD	0.10	
Y_{Zbam}	mg COD/mg COD	0.10	
Y_{Zbhm}	mg COD/mg COD	010	

3.3.4 Data Validation

BioWin is a comprehensive simulation platform that facilitates the ensemble modeling of wastewater treatment processes in WWTPs. It has been developed and widely used for engineering applications over more than a decade, and the coupling relationships between different modules, including the kinetic and stoichiometric parameters substituted in the present work, have been calibrated based on a large amount of studies and applications in the engineering practice. However, considering that model validation is a key step in determining whether the model can objectively describe and reflect the study system and whether the calculation results are reliable, six alternative process scenarios described in Sect. 3.3.2 were used as subjects for the calculation of the respective chemical consumption, GHG emissions, and energy consumption values using three sets of error criteria. By comparing these values, the model consistency and the reliability of the obtained data can be preliminarily determined. The results of the modeling calculations for S1 have a lower consistency, as shown in Table 3.5. However, in general, the relative errors between the output values of various scenarios are less than 1% for each error criterion. This indicates that the model calculation results are highly consistent. Therefore, to reduce the time used for model value calculations, 10^{-3} was used as the error criterion for the simulation platform.

Table 3.5 Summary of unique solution analysis results

Discharge standard	Error criterion	Chemical consumption		GHG emissions (kg CO_2-e/m^3)			Energy consumption (kWh/m^3)
		Ferric salts (t/d)	Methanol (10^3 m^3/d)	CH_4	N_2O	CO_2	
Class 2	10^{-3}	0.000	0.000	0.263	0.290	0.294	0.358
	10^{-4}	0.000	0.000	0.264	0.290	0.294	0.358
	10^{-5}	0.000	0.000	0.264	0.290	0.294	0.358
Class 1B	10^{-3}	0.090	0.000	0.503	0.295	0.355	0.428
	10^{-4}	0.089	0.000	0.502	0.295	0.355	0.428
	10^{-5}	0.090	0.000	0.503	0.295	0.355	0.428
Class 1A	10^{-3}	0.100	0.250	0.500	0.285	0.379	0.457
	10^{-4}	0.100	0.250	0.501	0.285	0.379	0.457
	10^{-5}	0.101	0.250	0.501	0.285	0.379	0.457

Adapted from Wang et al. (2012), with permission from Elsevier, Copyright (2012)

3.4 Environmental Impact Inventory Accounting and Analysis

3.4.1 Inventory Estimation Models

3.4.1.1 Energy Consumption

The estimated energy consumption (p_{ener}) mainly includes the energy consumed by the aeration system in the aerobic unit, RAS pump, MLR pump, mixer in the anaerobic (anoxic) unit, and heat consumption of the anaerobic sludge digestion unit. The energy consumption was calculated using the following equation:

$$p_{ener} = \frac{p_1 + p_2 + p_3 + p_4 + p_5 + p_6}{Q}, \tag{3.1}$$

where p_{ener} is the total energy consumption of the wastewater treatment process (kWh/m^3); p_1 is the energy consumption of the aeration system in the aerobic unit (kWh); p_2 is the energy consumption of the RAS pump (kWh); p_3 is the energy consumption of the MLR pump (kWh); p_4 is the energy consumption of the mixer in the anaerobic (anoxic) unit (kWh); p_5 is the energy consumption due to heat loss from the anaerobic sludge digestion unit (kWh); p_6 is the energy consumption in other parts of the WWTP, calculated based on 20% of the total energy consumption; and Q is the daily processing capacity of the WWTP (m^3/d).

The energy consumption of the aeration system in the aerobic unit (p_1) was calculated using Eq. 3.2:

$$p_1 = v_{\text{oxygen}} \cdot \eta_{\text{oxygen}}, \tag{3.2}$$

where v_{oxygen} is the oxygen transfer rate of the aerobic unit (kg O_2/d) obtained using the BioWin simulation and η_{oxygen} is the oxygen supply efficiency of the aerobic unit (kg O_2/kWh), which was set as 2 kg O_2/kWh in the model.

The energy consumption of the RAS or MLR pumps (p_2 or p_3) was calculated using the following equation:

$$p_{2(3)} = \frac{r \cdot Q_r \cdot h \cdot t_{\text{pump}}}{\eta_1 \cdot \eta_2} \times 10^3, \tag{3.3}$$

where r is the specific gravity of the sludge or mixed liquor (N/m^3); Q_r is the average flow rate of the sludge or mixed liquor (m^3/s); H is the elevation of the sludge or mixed liquor (m); η_1 is the pump efficiency, ranging from 0.65 to 0.85; η_2 is the motor efficiency, which is typically set to 0.95; and t_{pump} is the operating time of the RAS or MLR pumps (h).

The energy consumption of the mixer in the anaerobic (anoxic) unit (p_4) was calculated using Eq. (3.4):

$$p_4 = \frac{w_{\text{stir}} \cdot V_{\text{reactor}} \cdot t_{\text{stir}}}{1000}, \tag{3.4}$$

where w_{stir} is the mixer efficiency (W/m^3), which was set to 5 W/m^3 in the model; V_{reactor} is the operating volume of the tank (m^3); and t_{stir} is the operating time of the mixer (h).

The unit energy consumption due to heat loss from the anaerobic sludge digestion unit (p_5) was calculated using the following equation (Metcalf and Eddy 2003):

$$p_5 = \left\{ [V \cdot (T_\text{d} - T_\text{s}) \cdot t] \times \frac{4186.8}{24 \times 3600} + \sum \frac{F \cdot K \cdot (T_\text{d} - T_\text{a}) \cdot t}{3600} \times 1.2 \right\} \times 1.1, \tag{3.5}$$

where V is the volume of raw sludge fed into the digester per day (m^3/d); T_d is the digestion temperature (°C), which was set to 35 °C in the model; T_s is the original temperature of the raw sludge (°C); F is the heat dissipation area of the digester cover, walls, and bottom (m^2); T_a is the temperature of the medium (air or soil) outside the digester (°C); and K is the heat transfer coefficient of the digester cover, walls, and bottom [kJ/(m^2 h °C)], calculated using the following equation:

$$K = \frac{1}{\frac{1}{\alpha_1} + \sum \frac{\delta}{\lambda} + \frac{1}{\alpha_2}}, \tag{3.6}$$

where α_1 is the heat transfer coefficient of the internal wall of the digester, which is 1256 kJ/(m^2 h °C) for heat transfer from sludge to a reinforced concrete digester wall and 31.4 kJ/(m^2 h °C) for heat transfer from biogas to a reinforced concrete digester wall, and α_2 is the heat transfer coefficient of the external wall of the digester, that is, the heat transfer coefficient from the digester wall to the external medium. If the external medium is the atmosphere, the values range from 12.6 to 33.5 kJ/(m^2 h °C); if the external medium is soil, the values range from 2.1 to 6.3 kJ/(m^2 h °C); δ is the thickness of the structural and insulation layers of the digester (m); λ is the heat transfer coefficient of the structural and insulation layers of the digester, which is 5.6 kJ/(m^2 h °C) for reinforced concrete.

3.4.1.2 Greenhouse Gas (GHG) Emissions

For the analysis of GHG emissions (p_{gree}), the emissions of three types of gases, that is, methane (CH_4), nitrous oxide (N_2O), and carbon dioxide (CO_2), were considered. The amount of CH_4 was determined from the mass balance and transfer model in BioWin, N_2O was mainly accounted for based on the emission factors provided by the Intergovernmental Panel on Climate Change (IPCC) and data from the relevant key literature (see Table 3.6), while the energy consumption factors were mainly considered when determining the CO_2 production (see Table 3.6 for the estimation coefficients). In addition, the emissions of three types of GHGs were normalized for the calculation of the global warming potential (GWP), which is expressed as carbon dioxide equivalent (CO_2-eq). To obtain assessment results that are consistent with the national GHG inventory under the United Nations Framework Convention on Climate Change, the present work adopted the GWPs with a 100-year time horizon reported in the Third Assessment Report of the IPCC (2001), that is, GWPs of 25 kg CO_2-eq/kg and 298 kg CO_2-eq/kg for CH_4 and N_2O, respectively.

3.4.1.3 Chemical Consumption

In the assessment system of the present work, the chemical consumption (p_{chem}), including the exogenous carbon source (methanol, 10^3 m^3/d) and phosphorus-removing agent (ferric salts, t/d), was obtained from the BioWin simulation results.

3.4.2 Analysis of Chemical Consumption

As shown in Fig. 3.9, the average methanol consumption was $(0.09 \pm 0.05) \times 10^3$ m^3/d based on the Class 2 discharge standard. When the wastewater discharge standard was raised to Class 1B, the average methanol consumption increased to $(0.12 \pm 0.06) \times 10^3$ m^3/d, while the average ferric chloride consumption increased from 0 to 0.23 ± 0.12 t/d. When the discharge standard was raised to Class 1A, the average methanol consumption further increased to $(0.14 \pm 0.05) \times 10^3$ m^3/d, while the average ferric chloride consumption increased to 0.56 ± 0.26 t/d.

Table 3.6 Factors accounting for GHG emissions

Inventory	Lowest value	Median value	Highest value	Source
CH_4 produced from the degradation of organic matter in the effluent (kg CH_4/kg BOD)	0	0.06	0.12	IPCC (2006b)
Soluble CH_4 released from the effluent (kg CH_4)	–	Model value	–	BioWin
CH_4 produced during anaerobic sludge digestion (kg CH_4)	–	Model value	–	BioWin
Soluble CH_4 released from excess sludge (kg CH_4)	–	Model value	–	BioWin
N_2O released during biological treatment (kg N_2O–N/kg N)	0.003	0.035	0.253	Foley et al. (2010a, b)
N_2O released from the effluent (kg N_2O–N/kg N)	0.0005	0.0025	0.005	IPCC (1997, 2006a)
N_2O directly released from excess sludge (kg N_2O–N/kg N)	0.003	0.01	0.03	Doka (2003), IPCC (2006a)
CO_2 produced from CH_4 combustion for energy generation (g CO_2-e/kWh)	–	353	–	IPCC (2001)
CO_2 produced during electricity consumption (g CO_2-e/kWh)	–	877	–	IPCC (2001)

Note Median values were selected for case study calculations

Based on these results, wastewater discharge standards have a significant impact on the chemical consumption during the wastewater treatment process. When the standard was raised, the chemical consumption increased. In particular, an increase in the consumption of chemicals investigated in the present work, that is, the phosphorus-removing agent and exogenous carbon source, may lead to negative environmental impacts because of the following reasons: When the effluent from WWTPs is discharged to receiving waters, the salinity of the receiving waters increases due to the presence of residual salt-based chemicals in the wastewater effluent. In addition, during the wastewater treatment process, chemical agents used for phosphorus removal will increase the excess sludge production. Other excess chemical agents

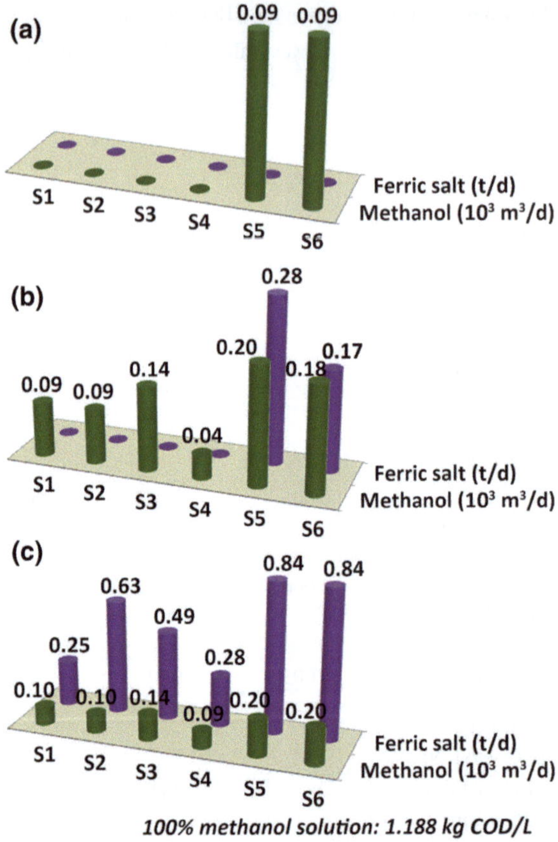

Fig. 3.9 Effects of discharge standards on the chemical consumption associated with alternative wastewater treatment scenarios: **a** Class 2; **b** Class 1B; **c** Class 1A. Adapted from Wang et al. (2012), with permission from Elsevier, Copyright (2012)

are also transferred to the excess sludge, thereby leading to an increase in the sludge treatment and disposal costs, such as energy consumption costs and transportation costs, which results in indirect environmental burdens.

In addition, compared with conventional A^2/O processes (scenarios S1–S4), reversed A^2/O processes (scenarios S5 and S6) result in a higher chemical consumption during the wastewater treatment process. This may be due to the early utilization of a portion of the organic carbon sources in the influent by bacteria in the anoxic unit in S5 and S6, which hinders the biological phosphorus release in the anaerobic unit to a certain extent due to the inadequacy of carbon sources. Therefore, to strictly meet the effluent standard for phosphorus, carbon source dosage and chemical-assisted phosphorus removal measures must be adopted in reversed A^2/O processes, which leads to disadvantages of such processes with respect to the reduction of the chemical consumption. Furthermore, compared with S5, S6 employs a step-feed mode, whereby 30% of the organic carbon sources in the influent directly enter the anaerobic unit and 70% of the organic carbon sources enter the anoxic unit.

However, the results indicate that the chemical consumption during the wastewater treatment process is not effectively reduced based on this approach.

Figure 3.9 shows that an increase in the wastewater discharge standard provides S4 with a relative advantage over other conventional A^2/O processes with respect to the reduction of the chemical consumption. In S4, RAS with a higher nitrate concentration is first returned to the anoxic unit for denitrification. Subsequently, sludge or mixed liquor with a small amount of nitrates is returned to the anaerobic unit. Therefore, the biological phosphorus release process in the anaerobic unit is not inhibited by the competition for carbon sources. From the perspective of chemical consumption reduction, S4 provides positive environmental effects because it mitigates eutrophication in receiving waters and reduces the negative environmental impact related to the substantial consumption of chemicals.

3.4.3 Analysis of Energy Consumption

As shown in Fig. 3.10, when the wastewater discharge standard was raised from Class 2 to Class 1A, the energy consumption required for the wastewater treatment process exhibited a significant upward trend (Class 2: 0.377 ± 0.023 kWh/m^3, Class 1B: 0.490 ± 0.089 kWh/m^3, Class 1A: 0.510 ± 0.065 kWh/m^3). From the perspective of fossil fuel depletion, high energy consumption during the wastewater treatment process and the increase of the consumption will lead to severe environmental burdens.

In general, the bulk energy consumption (46.9–58.5% of the total power consumption) is contributed by aeration in the aerobic unit of the wastewater treatment process. Therefore, methods to enhance the oxygen transfer efficiency and optimize the oxygen transfer process in the aerobic unit are keys to reduce the energy consumption during wastewater treatment. In addition, with increasingly stringent demands on the effluent quality, especially stricter limits of the phosphorus content, chemical methods are usually required to enhance the phosphorus removal during biological denitrification and phosphorus removal processes related to conventional wastewater treatment. However, such methods inevitably result in the generation of large amounts of excess sludge. To determine the relationship between energy consumption and excess sludge production, correlation analysis was performed. The results are shown in Fig. 3.11. It is apparent that a positive correlation exists between energy consumption and excess sludge production. This is because the energy consumed by heating and insulation during the anaerobic sludge digestion process increases when the amount of sludge increases. Therefore, a decrease in the sludge production will effectively reduce the energy consumption involved in sludge treatment and disposal. However, this is difficult to achieve during conventional wastewater treatment processes because the production of large amounts of sludge is required to achieve biological phosphorus removal. Based on the comparison of the energy consumption values of the various processes in Fig. 3.10, scenarios S5 and S6 are disadvantageous for energy consumption reduction, which may be due to the fact

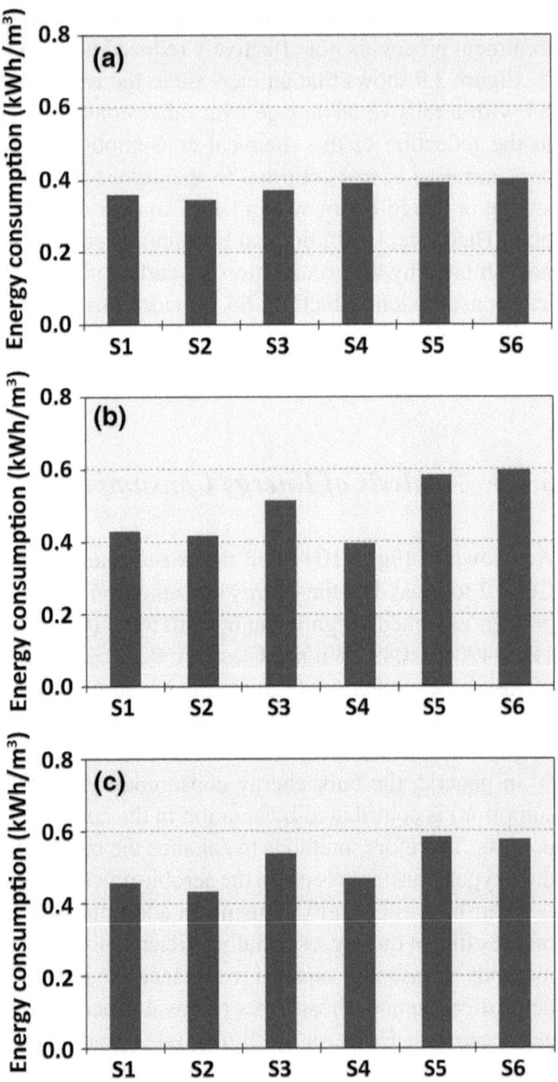

Fig. 3.10 Effects of discharge standards on the energy consumption associated with alternative wastewater treatment scenarios: **a** Class 2; **b** Class 1B; **c** Class 1A. Adapted from Wang et al. (2012), with permission from Elsevier, Copyright (2012)

that these two scenarios require the generation of substantial amounts of sludge to achieve phosphorus removal. This indicates that the enhancement of the phosphorus removal not only depends on chemical methods but is also achieved at the expense of a higher energy consumption. In addition, compared with other A^2/O type processes, S3 is also disadvantageous for energy consumption reduction. This may be because multiple points of activated sludge return were adopted in S3, which led to a higher energy consumption in the RAS pump. Compared with chemical consumption, a

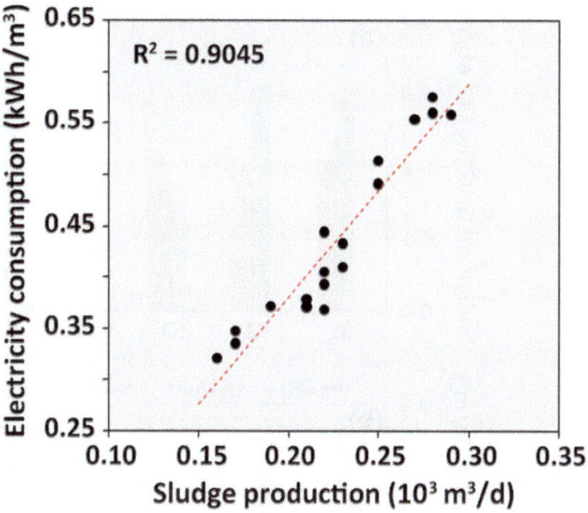

Fig. 3.11 Relationship between sludge production and electricity consumption. Adapted from Wang et al. (2012), with permission from Elsevier, Copyright (2012)

closer relationship exists between energy consumption and wastewater discharge standards; that is, stringent discharge standards are met at the expense of high energy consumption.

3.4.4 Analysis of GHG Emissions

Figure 3.12 shows the GHG emissions (CH_4, N_2O, CO_2, and total GHG) of various biological processes under different wastewater treatment standards. During the wastewater treatment process, most biodegradable organic compounds are converted to CO_2. However, this process has not been included in the GHG accounting by the IPCC. It is widely known that the release of CH_4 gas is closely associated with processes such as effluent discharge, wastewater treatment, and anaerobic sludge digestion. In the present study, it was assumed that the CH_4 released from the anaerobic digester can be used for the recovery of biological energy. The analysis of the results reveals that this measure reduces the GHG emissions of the anaerobic digester, thereby reducing the CH_4 contribution to 1.12–1.41%. Therefore, if the recovery of CH_4 generated from anaerobic sludge digestion during conventional wastewater treatment is considered, a win-win situation may be achieved with respect to the reduction of emissions and climate change mitigation. The process model adopted in the present work can be used to predict the amount of methane gas generated during wastewater treatment and blown off by the aeration system in the aerator unit. However, the CH_4 contribution is negligible compared with the amount of CH_4 produced during anaerobic sludge digestion. For N_2O, although a certain degree of uncertainty exists when accounting for N_2O generated during wastewater treatment due to limitations of the study conditions, the results of the present study indicate that stringent

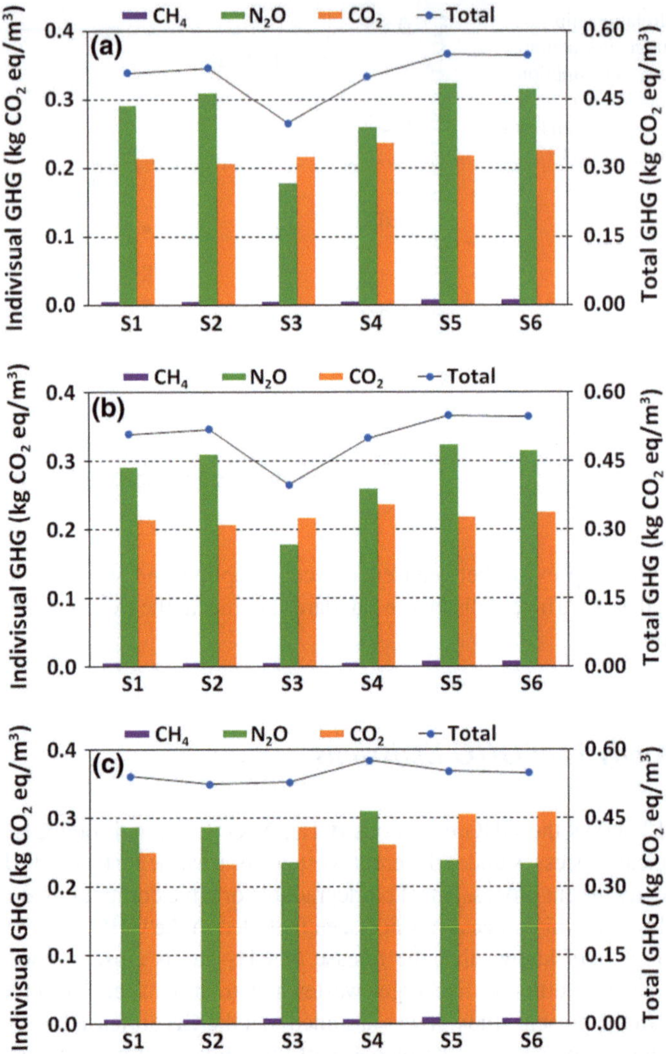

Fig. 3.12 Effects of discharge standards on the GHG emissions associated with alternative wastewater treatment scenarios: **a** Class 2; **b** Class 1B; **c** Class 1A. Adapted from Wang et al. (2012), with permission from Elsevier, Copyright (2012)

wastewater discharge standards lead to lower N_2O emission factors compared with relatively lenient discharge standards (Class 1A: 0.264 ± 0.033 kg CO_2-e/m^3, Class 1B: 0.269 ± 0.028 kg CO_2-e/m^3, Class 2: 0.279 ± 0.055 kg CO_2-e/m^3). Such a conclusion has also been supported by reports from other researchers (Foley et al. 2010b). The amount of CO_2 released is positively correlated with the increasingly stringent wastewater treatment standards (Class 2: 0.310 ± 0.020 kg CO_2-e/m^3, Class 1B: 0.409 ± 0.078 kg CO_2-e/m^3, Class 1A: 0.426 ± 0.057 kg CO_2-e/m^3). Such a pattern is similar to that of energy consumption because CO_2 release mainly arises from energy consumption. In general, the total GHG emissions exhibit an increasing trend with increasingly stringent wastewater discharge standards (Class 2: 0.970 ± 0.219 kg CO_2-e/m^3, Class 1B: 1.298 ± 0.171 kg CO_2-e/m^3, Class 1A: 1.307 ± 0.269 kg CO_2-e/m^3). From the perspective of GHG emissions, S3 is the most suitable choice for wastewater treatment.

3.5 Summary

This chapter discussed the adoption of LCA for an environmental impact inventory analysis of six typical wastewater treatment scenarios from the perspectives of chemical consumption, energy consumption, and GHG emission inventory. On this basis, the impact of wastewater discharge standards on environmental burdens created by wastewater treatment was investigated. The results are as follows:

(1) Substantial differences exist in the chemical consumption, energy consumption, and GHG emission inventory of the six wastewater treatment processes. Process S4 is the optimal choice for wastewater treatment from the perspective of the reduction of the chemical consumption, S2 is the most ideal choice for energy consumption reduction, while S3 is the best choice for the reduction of GHG emissions.

(2) When the wastewater discharge standard was raised from Class 2 to Class 1B and to Class 1A, the average consumption of the exogenous carbon source (methanol) was 0.09×10^3, 0.12×10^3, and 0.14×10^3 m^3/d, respectively; the average consumption of the chemical phosphorus-removing agent (ferric chloride) was 0, 0.23, and 0.56 t/d, respectively; the average energy consumption was 0.377, 0.490, and 0.510 kWh/m^3, respectively; and the average total GHG emissions were 0.970, 1.298, and 1.307 kg CO_2-e/m^3, respectively.

(3) From the perspective of life cycle environmental impacts, as the discharge standard for contaminants in wastewater becomes increasingly stringent, a good effluent quality is achieved at the expense of increased chemical consumption, energy consumption, and GHG emissions. Although stringent wastewater discharge standards ensure the quality of water environments in receiving waters, they also increase the complexity of wastewater treatment processes and aggravate negative environmental impacts generated by wastewater treatment.

References

Bagley, D. M., & Brodkorb, T. S. (1999). Modeling microbial kinetics in an anaerobic sequencing batch reactor—Model development and experimental validation. *Water Environment Research, 71*(7), 1320–1332.

Barker, P. S., & Dold, P. L. (1997a). General model for biological nutrient removal activated-sludge systems: Model application. *Water Environment Research, 69*(5), 985–991.

Barker, P. S., & Dold, P. L. (1997b). General model for biological nutrient removal activated-sludge systems: Model presentation. *Water Environment Research, 69*(5), 969–984.

Costello, D. J., Greenfield, P. F., & Lee, P. L. (1991). Dynamic modeling of a single-stage high-rate anaerobic reactor. 1. Model derivation. *Water Research, 25*(7), 847–858.

Doka, G. (2003) *Life cycle inventories of waste treatment services* (Ecoinvent Report No. 13). Dubendorf: Swiss Centre for Life Inventories.

Dold, P. L., Ekama, G. A., & Marais, G. R. (1980). The activated sludge process. 1. A General Model For The Activated Sludge Process. *Progress in Water Technology, 12*(6), 47–77.

Eldyasti, A., Nakhla, G., & Zhu, J. (2012). Development of a calibration protocol and identification of the most sensitive parameters for the particulate biofilm models used in biological wastewater treatment. *Bioresource Technology, 111,* 111–121.

Envirosim. (2007) BioWin process simulator. Envirosim Associates Ltd.

Foley, J., de Haas, D., Hartley, K., & Lant, P. (2010a). Comprehensive life cycle inventories of alternative wastewater treatment systems. *Water Research, 44*(5), 1654–1666.

Foley, J., de Haas, D., Yuan, Z. G., & Lant, P. (2010b). Nitrous oxide generation in full-scale biological nutrient removal wastewater treatment plants. *Water Research, 44*(3), 831–844.

Guinee, J. B. (2002). *Handbook on life cycle assessment: Operational guide to the ISO standards.* Springer.

Hafez, H., Naggar, M. H. E., & Nakhla, G. (2010). Steady-state and dynamic modeling of biohydrogen production in an integrated biohydrogen reactor clarifier system. *International Journal of Hydrogen Energy, 35*(13), 6634–6645.

Horne, R., Grant, T., & Verghese, K. (2009). *Life cycle assessment: principles, practice and prospects.* VIC, Australia: CSIRO Publishing.

IPCC. (1997) *Reference manual: Intergovenmental panel on climate change.*

IPCC. (2001). *Climate change 2001: The scientific basis.* Cambridge: Cambridge University Press.

IPCC. (2006a). IPCC guidelines for national greenhouse gas inventories. In *National greenhouse gas inventories programme.*

IPCC. (2006b). Wastewater treatment and discharge. H. S. Eggleston, L. Buendia, K. Miwa, T. Ngara, & K. Tanabe (Eds.), *The national greenhouse gas inventories programme.* Japan.

Jones, R., Parker, W., Zhu, H., Houweling, D., & Murthy, S. (2009). Predicting the degradability of waste activated sludge. *Water Environment Research, 81*(8), 765–771.

Kang, X. S., Liu, C. Q., Zhang, B., Bi, X. J., Zhang, F., & Cheng, L. H. (2011). Application of reversed A(2)/O process on removing nitrogen and phosphorus from municipal wastewater in China. *Water Science and Technology, 63*(10), 2138–2142.

Liu, W. J., Hu, Z. R., Walker, R. L., & Dold, P. L. (2011). Enhanced nutrient removal MBR system with chemical addition for low effluent TP. *Water Science and Technology, 64*(6), 1298–1306.

Liwarska-Bizukojc, E., & Biernacki, R. (2010). Identification of the most sensitive parameters in the activated sludge model implemented in BioWin software. *Bioresource Technology, 101*(19), 7278–7285.

Liwarska-Bizukojc, E., Olejnik, D., Biernacki, R., & Ledakowicz, S. (2011). Calibration of a complex activated sludge model for the full-scale wastewater treatment plant. *Bioprocess and Biosystems Engineering, 34*(6), 659–670.

Lundie, S., Peters, G. M., & Beavis, P. C. (2004). Life Cycle Assessment for sustainable metropolitan water systems planning. *Environmental Science and Technology, 38*(13), 3465–3473.

Makinia, J. (2010). *Mathematical modelling and computer simulation of activated sludge systems.* IWA Publishing.

Masse, D. I., & Droste, R. L. (2000). Comprehensive model of anaerobic digestion of swine manure slurry in a sequencing batch reactor. *Water Research, 34*(12), 3087–3106.

McKinney, R. E. (1960). Complete mixing activated sludge. *Water and Sewage Works, 107*(2), 69.

Metcalf, I., & Eddy, H. (2003). *Wastewater engineering: Treatment and reuse*. New York: McGraw-Hill.

Mosey, F. E. (1983). Mathematical-modeling of the anaerobic-digestion process—Regulatory mechanisms for the formation of short-chain volatile acids from glucose. *Water Science and Technology, 15*(8–9), 209–232.

Mulkerrins, D., Jordan, C., McMahon, S., & Colleran, E. (2000). Evaluation of the parameters affecting nitrogen and phosphorus removal in anaerobic/anoxic/oxic (A/A/O) biological nutrient removal systems. *Journal of Chemical Technology and Biotechnology, 75*(4), 261–268.

Pasztor, I., Thury, P., & Pulai, J. (2009). Chemical oxygen demand fractions of municipal wastewater for modeling of wastewater treatment. *International Journal of Environmental Science and Technology, 6*(1), 51–56.

Shaw, A., Kadava, A., & Tarallo, S. (2011). *Refinement of Life Cycle Assessment (LCA) methods for water and wastewater treatment plant design*. Amsterdam: IWA publisher.

Tillman, A.-M., Svingby, M., & Lundström, H. (1998). Life cycle assessment of municipal waste water systems. *International Journal of Life Cycle Assessment, 3*(3), 145–157.

Wang, X., Liu, J. X., Ren, N. Q., & Duan, Z. S. (2012). Environmental profile of typical anaerobic/anoxic/oxic wastewater treatment systems meeting increasingly stringent treatment standards from a life cycle perspective. *Bioresource Technology, 126,* 31–40.

Wang, J. H., Zhang, J., Xie, H. J., Qi, P. Y., Ren, Y. G., & Hu, Z. (2011). Methane emissions from a full-scale A/A/O wastewater treatment plant. *Bioresource Technology, 102*(9), 5479–5485.

Zeng, W., Li, L., Yang, Y. Y., Wang, S. Y., & Peng, Y. Z. (2010). Nitritation and denitritation of domestic wastewater using a continuous anaerobic-anoxic-aerobic (A(2)O) process at ambient temperatures. *Bioresource Technology, 101*(21), 8074–8082.

Zhou, Z., Wu, Z. C., Wang, Z. W., Tang, S. J., Gu, G. W., Wang, L. C., et al. (2011). Simulation and performance evaluation of the anoxic/anaerobic/aerobic process for biological nutrient removal. *Korean Journal of Chemical Engineering, 28*(5), 1233–1240.

Chapter 4
A Refined Assessment Methodology for Wastewater Treatment Alternatives

4.1 Overview

The environmental quality of receiving waters will no longer be the sole objective of sustained management in future wastewater treatment. In addition to the protection of water resources and environments, considerable attention must be paid to other resources, such as energy and nutrient resources, for long-term sustained development. It is certain that the reduction of the energy consumption and greenhouse gas (GHG) emissions and resource recovery will become key focus areas in the development of new wastewater treatment technologies and processes in the future. The exploration of methods to construct novel ensemble-type wastewater treatment technologies and processes aiming at energy conservation, reduced carbon emissions, and resource recovery based on existing technologies and processes is the future trend in the development of the wastewater industry. As described in Sect. 2.3, comprehensive assessment systems of wastewater treatment processes were established based on the assurance of environmental benefits such as the water quality. In Chap. 3, the sources and key links of the environmental impact of typical wastewater treatment processes were preliminarily analyzed from the life cycle perspective. However, there is a lack of consideration of the recovery and utilization of usable materials in wastewater and sludge and relevant scenario settings and analysis. Therefore, regardless of the research and development of new technologies and new processes in the future or the upgrade and reconstruction of existing processes, a scientific, systematic, and comprehensive wastewater treatment assessment system is required. Such a system should consider the technological levels of existing wastewater treatment processes and the main environmental effects of the wastewater treatment and resource recovery potential of materials produced during the treatment.

Based on the aforementioned research background, the concept of multi-objective management of municipal wastewater treatment, with an emphasis on energy conservation, reduced carbon emissions, and resource recovery from pollutants, has been proposed in the first section of the present chapter. Subsequently, based on the envi-

X. Wang, *Energy Consumption, Chemical Use and Carbon Footprints of Wastewater Treatment Alternatives*, Springer Theses,
https://doi.org/10.1007/978-981-13-5983-5_4

ronmental impact inventory analysis performed in Chap. 3 and the minimization of the environmental footprint and maximization of resource recovery, the assessment system for the multi-objective management of municipal wastewater treatment was constructed including aspects such as boundary definition, data acquisition, data processing, and comprehensive analysis. In this system, four scenarios, including the environmental footprint and bioenergy recovery scenarios, were first proposed for scenario analysis based on the requirements for the sustained management of wastewater treatment and considering technological levels of existing wastewater and sludge treatment and disposal methods. Concurrently, a set of algorithms was established for the versatile and rapid processing of data and to overcome assessment barriers arising from different units and orders of magnitude of assessment indicators, assisting decision-makers in the rapid assessment and identification of optimum wastewater treatment schemes and process combinations. The research described above represents the main novelties of this chapter.

4.2 Proposal of the Concept of Multi-objective Management of Municipal Wastewater Treatment

Following the United Nations Conference on Environment and Development in 1992, countries around the world made tremendous progress in the formulation and implementation of strategies for sustainable development. The concept of sustainable development has gradually taken root and has become the guiding ideology behind the formulation of development strategies by the governments of various countries and regions. At this point in time, with a global emphasis on sustainable development strategies, a wide range of industries can contribute to sustainable development; the wastewater treatment industry is no exception. Currently, secondary biological processes that are widely applied for the treatment of municipal wastewater can effectively remove substances such as organic matter, nitrogen, and phosphorus from wastewater, thereby assuring the water quality of the receiving waters. However, a number of recent reports (Foley et al. 2010b; Shahabadi et al. 2009; Verstraete and Vlaeminck 2011) indicated that the development of secondary biological processes for wastewater treatment faces great challenges, mainly due to environmental problems such as high energy consumption, excess sludge production, and GHG emissions caused by the wastewater treatment process. Therefore, the effluent quality should not be the sole indicator of the sustained management of wastewater treatment. In addition to ensuring the sustainability of healthy water resources, attention must be paid to other environmental issues to enable a long-term sustained development. Otherwise, wastewater treatment will merely be a technical means of pollution transfer (Foley et al. 2010a) because the assurance of the quality of the water environment is achieved by shifting the pollution problem to atmospheric pollution in the form of GHG and solid waste pollution based on substantial amounts of excess sludge.

From the perspective of sustainability, wastewater can be viewed as a carrier of energy sources and resources rather than a mere pollutant or pollution source. In fact, organic matter, nitrogen, and phosphorus in wastewater and excess sludge generated in biological treatment units are usable and recoverable resources. For instance, sludge is primarily composed of microbes and organic material. Microbial cells are also organic matter, which can mainly be recovered through methanogenesis during the anaerobic fermentation of sludge, biological hydrogen production, and microbial fuel cells (MFCs) (Logan et al. 2006; Ren et al. 2011). Phosphorus in wastewater is also a potential recoverable phosphorus source; some researchers proved the possibility of phosphorus recovery in the form of struvite for agricultural use from supernatant generated during anaerobic sludge fermentation (Le Corre et al. 2009). Therefore, the focus of the wastewater treatment should be shifted from conventional removal and disposal methods to resource recovery management methods. In short, the energy and resources contained in wastewater and sludge should be recovered to the greatest extent possible; at the same time, low-energy and material consumption must also be achieved by wastewater treatment technologies.

In consideration of the aforementioned points, we propose the multi-objective management of municipal wastewater treatment, which involves the following: Under the premise that the effluent quality meets certain discharge standards, the energy and material consumption required for wastewater treatment are reduced, the environmental footprint (e.g., GHG emissions) is reduced, and existing recovery technologies are utilized for the recovery of organic matter and nutrients, such as nitrogen and phosphorus from wastewater and excess sludge, via multiple routes to manage multiple objectives such as energy conservation, reduced carbon emissions, and resource recovery. In other words, multi-objective management of municipal wastewater treatment has a twofold significance: the first being the satisfaction of certain wastewater treatment requirements and standards to ensure the sustainability of the health of receiving waters, which is the most fundamental and most important objective; the second being the minimization of environmental burdens and the environmental footprint generated during wastewater treatment and the maximization of the recovery of usable substances from wastewater and sludge, which is a continuous objective that should be particularly considered and implemented from now on. The proposal of the concept of multi-objective management of municipal wastewater treatment promotes the harmonious integration of wastewater treatment and overall ecological benefits, thereby facilitating the sustainable development of wastewater treatment.

4.3 Scenario Definition for Multi-objective Management

The term "scenario" was coined by Kahn and Wiener in 1967 and is defined as the description of a succession of events that leads to the development of a situation from the initial state to the future state. This includes qualitative and quantitative descriptions of the basic characteristics of various states and descriptions of the possibility of the occurrence of various states. Scenario analysis is a process that involves the prediction of the occurrence of these states and comparative analysis of possible impacts. A core step in the construction of the assessment framework and subsequent scenario analysis is the scenario definition. Therefore, the four below-mentioned scenarios have been built on the basis of existing technological levels of wastewater and sludge processing and disposal and prediction of the future direction of sustained management.

(1) Environmental footprint scenario

Based on the assumption that the environments in which wastewater treatment plants (WWTPs) are located will become increasingly sensitive in the future due to the impact of various factors, the quantification and assessment of the environmental footprint of WWTPs and formulation of effective counter-strategies are highly necessary. Therefore, environmental burdens are primarily considered in the environmental footprint scenario. The research approach in Chap. 3 (environmental impact inventory analysis of typical wastewater treatment processes) was used as reference for the definition and construction of this scenario. Energy consumption, chemical consumption, and GHG emissions were used to represent the environmental footprint inventory.

(2) Bioenergy recovery scenario

Based on the assumption of severe depletion of fossil fuels and the increasing prominence of the environmental impact of traditional fossil fuels (e.g., coal and petroleum), the self-supply or self-compensation of energy sources in WWTPs will become inevitable. The utilization of organic matter in wastewater or sludge and the transformation of such matter into carriers of energy sources form the core of the bioenergy recovery scenario. Therefore, biogas (primarily CH_4) generated during the anaerobic digestion of excess sludge is collected for energy recovery through biogas combustion in this scenario. It must be noted that the environmental footprint of the treatment and disposal of wastewater and sludge, including the energy recovery process, is also considered in this scenario, in addition to the potential of energy recovery.

(3) Nutrient recovery scenario

In this scenario, it is assumed that future WWTPs will produce a huge amount of excess sludge and pollution transfer problems due to sludge disposal methods, such as sludge incineration, will become increasingly severe. To increase the recovery and utilization rates of resources in waste material and to replenish the soil fertility, the excess sludge produced by the WWTP is thickened and dewatered before transportation out of the plant for nutrient recovery via natural composting methods. In addition, the following points must be noted: (1) The excess sludge mainly originates from municipal WWTPs; (2) the ecotoxic impacts of residual heavy metals in the excess sludge are neglected; (3) in addition to investigating the nitrogen and phosphorus contents of excess sludge, only the environmental footprint (energy consumption, chemical consumption, and GHG emissions) of the wastewater and sludge treatment and disposal is taken into consideration.

(4) Simultaneous bioenergy and nutrient recovery scenario

Assuming that the background problems of scenarios (1) and (3) simultaneously occur and become increasingly severe, the main purpose of this scenario is to enhance the bioenergy and nutrient recovery and management while considering methods to reduce the environmental footprint. In the simultaneous bioenergy and nutrient recovery scenario, CH_4 produced from excess sludge of the WWTP during the anaerobic digestion process will be collected for energy recovery via combustion; simultaneously, nitrogen and phosphorus in the supernatant produced during anaerobic digestion of the sludge are recovered in the form of struvite, while the anaerobically digested sludge is thickened, dewatered, and subsequently transported out of the plant for agricultural use via natural composting. The environmental footprint created by both wastewater treatment and resource recovery, including the resource recovery potential of pollutants, is considered in this scenario.

4.4 Basic Principles of the Construction of the Assessment System for Multi-objective Management

The construction of an assessment system heavily depends on the establishment of an indicator system. Therefore, it is vital to pay much attention while developing the indicator system, because it can specifically and objectively reflect the characteristics of the study system. Assessment indicators are defined as parameters used for the assessment of a system and are used to reflect the concept of quantity and specific values of certain key factors or phenomena in the system. The indicator system is composed of several interrelated and mutually complementary indicators and is used for the comprehensive, systematic, and accurate reflection of important properties of the studied subject.

Because municipal WWTPs are the subject of the present assessment system, which assesses the role of WWTPs in multi-objective management of wastewater treatment, this system is a complex multi-objective, multi-hierarchical, and multi-functional system. Therefore, the comprehensive assessment is impacted by multiple factors such as the effluent quality, energy consumption, chemical consumption, GHG emissions, and energy and resource recovery. The selection of assessment indicators based on the requirements of different scenarios is an exploration and creation process, that is, "building something from nothing," while the selection of several highly representative factors from a large number of assessment indicators represents the subsequent betterment process, that is, "selecting the best from something." Accordingly, the basic principles that must be followed during the construction of an assessment system for multi-objective management of wastewater treatment are the following:

(1) Practicality and scientificity

During the selection of assessment indicators, the scientificity and correctness of the indicator system must first be considered from a theoretical standpoint. In other words, the construction of the indicator system must be performed on a scientific basis and the selected indicators must objectively reflect the actual impact of the studied subject. Secondly, particular emphasis should be placed on the comparability and practicality of the indicators; that is, the indicators of the assessment system must not be randomly selected; instead, the degree to which actual problems can be reflected by the selected indicators during wastewater and sludge treatment must be considered based on the concept of multi-objective management of wastewater treatment to determine the practical value of each indicator.

(2) Comprehensiveness and hierarchicality

In this assessment system, wastewater is no longer regarded as waste matter that is discharged after treatment; instead, the organic material in wastewater, including the excess sludge produced during the treatment process, is regarded as a usable resource. Due to the involvement of multiple factors, the assessment system for multi-objective management of wastewater treatment is a comprehensive system. On the one hand, it must comprehensively reflect the overall benefits of sustained management of wastewater in the future; on the other hand, it must also reflect the main characteristics promoted by different scenarios, such as the environmental footprint, bioenergy recovery, and nutrient recovery scenarios, and mutual coordination and mutual enhancement between the scenarios. Because such a complex system is composed of subtarget layers within the various layers with different functions, hierarchicality should be present in the selection of assessment indicators. In other words, higher target layers should be the ultimate goals of lower target layers, while lower target layers represent subdivisions of higher target layers.

(3) Existence of principal components and independency

The principle of the existence of principal components is as follows: In accordance with the complexity of WWTPs, a certain number of key variables that can accurately represent the characteristics and behavior of the system are selected from a large number of factors based on the degree of importance and contribution to the target system. The number of selected indicators has an extremely important influence on the assessment results. If a small number of indicators are selected, the actual behavior of the system cannot be completely characterized and there is a higher possibility that key information is omitted; if a large number of indicators are used, the excessive amount of information may increase the difficulty and workload of the analysis process. This is known as the principle of independency. Therefore, when constructing the assessment system for the multi-objective management of wastewater treatment, attention should be paid to the representativeness and comprehensiveness of the selected indicators to strive toward a comprehensive assessment of the overall benefits of different wastewater management objectives.

(4) Versatility and operability

When defining the indicator system for the assessment of multi-objective management, versatile processing is required to satisfy the requirements of various scenarios. Different requirements are associated with different objectives; for instance, the environmental burdens created during wastewater treatment are the main concern of administrators. Therefore, corresponding assessments will focus on the overall impacts on the environment. By contrast, economic benefits are the main concern of wastewater management personnel. Therefore, greater emphasis will be placed on the energy recovery to achieve the self-supply or self-compensation of energy sources in WWTPs. Operability refers to the consideration of the degree of ease/difficulty of acquiring data, the reliability of the acquired data when selecting assessment indicators, and the assurance that the data can be processed by easy-to-learn mathematical methods to assist decision-makers in performing rapid and accurate assessments.

4.5 Construction of the Assessment Framework for Multi-objective Management of Wastewater Treatment

As shown in Fig. 4.1, the assessment framework must be established before performing the assessment of multi-objective management of wastewater treatment and obtaining the final assessment conclusions using the following four major steps: (1) delineation of assessment boundaries; (2) selection of the indicator system; (3) acquisition of assessment data; (4) characterization of assessment indicators.

4.5.1 Delineation of Assessment Boundaries

To facilitate effective assessments and enhance the accuracy of the assessment process and results, the delineation of assessment boundaries is required before the assessment of multi-objective management of wastewater treatment can be performed. Because the primary objective of wastewater treatment is to ensure that the effluent quality meets certain water quality requirements for the assurance of the quality of receiving waters, the premise for the assessment of multi-objective management is as follows: Wastewater containing excessive levels of pollutants enters the WWTP and is subjected to appropriate biological, physical, and chemical processes; consequently, the effluent meets relevant standards and satisfies the quality requirements for receiving waters. In the present assessment system, the overall objective is defined based on the overall benefits of municipal wastewater treatment. The start

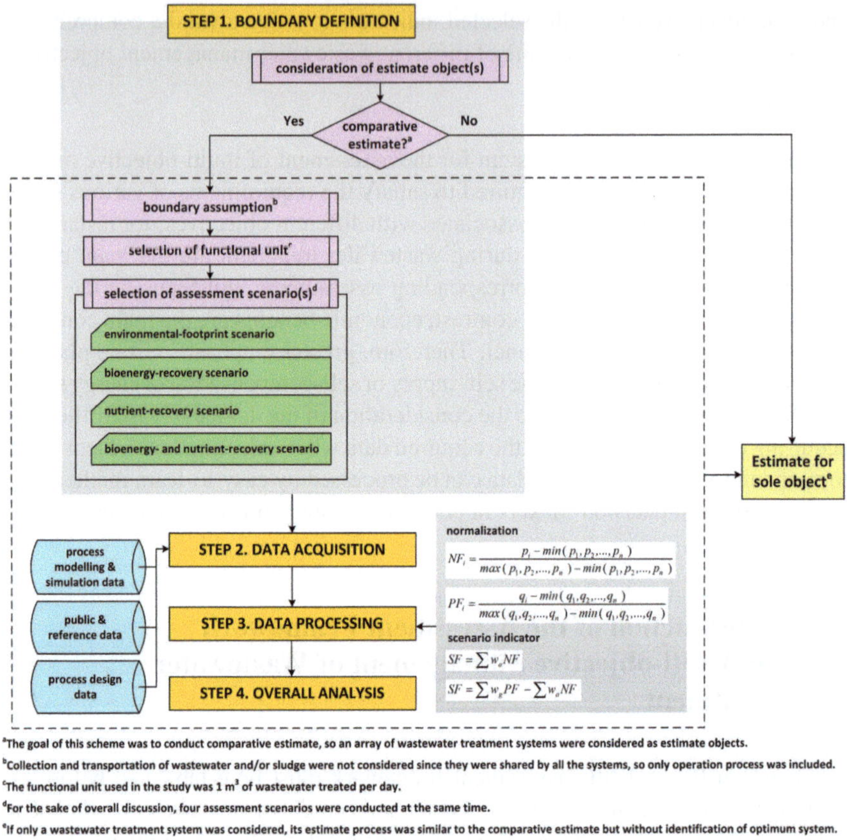

Fig. 4.1 Framework of the assessment system for multi-objective management of wastewater treatment. Adapted from Wang et al. (2012), with permission from American Chemical Society, Copyright (2012)

point of the assessment is the inflow of wastewater into the municipal WWTP, while the end point of the assessment is the discharge of a standard-compliant effluent and the treatment and disposal of sludge. Due to condition and time constraints, the consumption of raw materials during the construction and decommissioning stages and the resource consumption and environmental impacts arising from other processes have not been considered in the present assessment. Before conducting the assessment, the appropriate scenario should be selected from the four defined scenarios, that is, the environmental footprint scenario, bioenergy recovery scenario, nutrient recovery scenario, and simultaneous bioenergy and nutrient recovery scenario (refer to Sect. 4.3 for detailed descriptions of the four scenarios), based on actual assessment needs. The functional unit adopted in the assessment process is 1 m^3/d.

4.5.2 Selection of the Indicator System

The assessment indicator system of each scenario should include the required information to the greatest extent possible. For the environmental footprint scenario, the environmental impacts generated during wastewater treatment, such as climate change and abrupt atmospheric change, are the main focus areas of the assessment. Some researchers reported that the chemical consumption, energy consumption, and GHG emissions that result from wastewater treatment in municipal WWTPs are closely linked to adverse environmental impacts such as global warming, ozone depletion, acidification, and ecotoxicity (Foley et al. 2010a; Horne et al. 2009; Shaw et al. 2011). Therefore, energy consumption (p_{ener}), chemical consumption (p_{chem}), and GHG emissions (p_{gree}) generated during wastewater treatment form the primary assessment indicator system of the environmental footprint scenario. The respective indicator systems of the bioenergy and nutrient recovery scenarios include the amount of bioenergy recovered (q_{bioe}) and amount of nutrients recovered (q_{nitr} and q_{phos}), in addition to p_{ener}, p_{chem}, and p_{gree}. The indicator system for the simultaneous bioenergy and nutrient recovery scenario includes the amount of struvite produced (q_{stru}), in addition to p_{ener}, p_{chem}, p_{gree}, q_{bioe}, q_{nitr}, and q_{phos}.

4.5.3 Acquisition of Assessment Data

To reduce the uncertainties introduced by conventional data acquisition routes (see Sect. 2.3), the BioWin simulator (V.3.0., BW3-1952) employed in Chap. 3 was also used in this chapter for the acquisition of basic assessment data. For data that could not be acquired through numerical modeling, relevant data from public databases or key reference literature were used. The estimation methods and models for the indicators involved in the assessment system and their numerical values are described in the following subsections.

4.5.3.1 Energy Consumption

The estimated energy consumption (p_{ener}) mainly includes the energy consumed by the aeration system in the aerobic unit, return activated sludge (RAS) pump, mixed liquor recirculation (MLR) pump, mixer in the anaerobic (anoxic) unit, and unit heat consumption in the anaerobic sludge digestion unit. The energy consumption was calculated using the following equation:

$$p_{\mathrm{ener}} = \frac{p_1 + p_2 + p_3 + p_4 + p_5 + p_6}{Q}, \tag{4.1}$$

where p_{ener} is the total energy consumption of the wastewater treatment process (kWh/m^3); p_1 is the energy consumption of the aeration system in the aerobic unit (kWh); p_2 is the energy consumption of the RAS pump (kWh); p_3 is the energy consumption of the MLR pump (kWh); p_4 is the energy consumption of the mixer in the anaerobic (anoxic) unit (kWh); p_5 is the energy consumption due to heat loss from the anaerobic sludge digestion unit (kWh); p_6 is the energy consumption in other parts of the WWTP, calculated based on 20% of the total energy consumption; and Q is the daily processing capacity of the WWTP (m^3/d).

The energy consumption of the aeration system in the aerobic unit (p_1) was calculated using Eq. (4.2):

$$p_1 = v_{\text{oxygen}} \cdot \eta_{\text{oxygen}}, \tag{4.2}$$

where v_{oxygen} is the oxygen transfer rate of the aerobic unit (kg O_2/d) obtained from the BioWin simulation and η_{oxygen} is the oxygen supply efficiency of the aerobic unit (kg O_2/kWh), which was set to 2 kg O_2/kWh in the model.

The energy consumption of the RAS or MLR pumps (p_2 or p_3) was calculated using the following equation:

$$p_{2(3)} = \frac{r \cdot Q_r \cdot h \cdot t_{\text{pump}}}{\eta_1 \cdot \eta_2} \times 10^3, \tag{4.3}$$

where r is the specific gravity of the sludge or mixed liquor (N/m^3); Q_r is the average flow rate of the sludge or mixed liquor (m^3/s); h is the elevation of the sludge or mixed liquor (m); η_1 is the pump efficiency, ranging from 0.65 to 0.85; η_2 is the motor efficiency, which is typically set to 0.95; and t_{pump} is the operating time of the RAS or MLR pumps (h).

The energy consumption of the mixer in the anaerobic (anoxic) unit (p_4) was calculated using Eq. (4.4):

$$p_4 = \frac{w_{\text{stir}} \cdot V_{\text{reactor}} \cdot t_{\text{stir}}}{1000}, \tag{4.4}$$

where w_{stir} is the mixer efficiency (W/m^3), which was set to 5 W/m^3 in the model; V_{reactor} is the operating volume of the tank (m^3); and t_{stir} is the operating time of the mixer (h).

The unit energy consumption due to heat loss from the anaerobic sludge digestion unit (p_5) was calculated using the following equation (Metcalf and Eddy 2003):

$$p_5 = \left\{ [V \cdot (T_d - T_s) \cdot t] \times \frac{4186.8}{24 \times 3600} + \sum \frac{F \cdot K \cdot (T_d - T_a) \cdot t}{3600} \times 1.2 \right\} \times 1.1, \tag{4.5}$$

where V is the volume of the raw sludge fed into the digester per day (m^3/d); T_d is the digestion temperature (°C), which was set to 35 °C in the model; T_s is the original temperature of the raw sludge (°C); F is the heat dissipation area of the digester cover, walls, and bottom (m^2); T_a is the temperature of the medium (air or soil) outside the digester (°C); and K is the heat transfer coefficient of the digester cover, walls, and bottom [kJ/(m^2 h °C)] calculated using the following equation:

$$K = \frac{1}{\frac{1}{\alpha_1} + \sum \frac{\delta}{\lambda} + \frac{1}{\alpha_2}}, \quad (4.6)$$

where α_1 is the heat transfer coefficient of the internal wall of the digester, which is 1256 kJ/(m^2 h °C) for the heat transfer from sludge to a reinforced concrete digester wall and 31.4 kJ/(m^2 h °C) for the heat transfer from biogas to a reinforced concrete digester wall, and α_2 is the heat transfer coefficient of the external wall of the digester, that is, the heat transfer coefficient from the digester wall to the external medium. If the external medium is atmosphere, the values lie within the range of 12.6–33.5 kJ/(m^2 h °C); if the external medium is soil, the values lie within the range of 2.1–6.3 kJ/(m^2 h °C). The parameter δ is the thickness of the structural and insulation layers of the digester (m) and λ is the heat transfer coefficient of the structural and insulation layers of the digester, which is 5.6 kJ/(m^2 h °C) for reinforced concrete.

4.5.3.2 Chemical Consumption

In the assessment system of the present work, the consumption of chemicals (p_{chem}), including the exogenous carbon source (methanol, 10^3 m^3/d) and phosphorus removing agent (ferric salts, t/d), was obtained from the BioWin simulation results.

4.5.3.3 GHG Emissions

For the analysis of GHG emissions (p_{gree}), the emissions of three types of gases, that is, methane (CH_4), nitrous oxide (N_2O), and carbon dioxide (CO_2), were mainly considered. The amount of CH_4 was determined from the mass balance and transfer model in BioWin, N_2O was mainly accounted for using the emissions factors provided by the Intergovernmental Panel on Climate Change (IPCC) and data from relevant key literature (see Table 4.1), while the energy consumption factors were mainly considered when accounting for the CO_2 production (see Table 4.1 for the estimation coefficients). In addition, the emissions of the three types of GHGs were normalized for the calculation of the global warming potential (GWP), which is expressed in terms of the carbon dioxide equivalent (CO_2-eq). To obtain assessment results that are consistent with the national GHG inventory under the United Nations Framework Convention on Climate Change, the present work adopted GWPs with a 100-year time horizon reported in the Third Assessment Report of the IPCC (2001), that is, GWPs of 25 kg CO_2-eq/kg and 298 kg CO_2-eq/kg for CH_4 and N_2O, respectively.

Table 4.1 Factors accounting for greenhouse gas emissions

Inventory	Lowest value	Median value	Highest value	Source
CH_4 produced from the degradation of organic matter in the effluent (kg CH_4/kg BOD)	0	0.06	0.12	IPCC (2006b)
Soluble CH_4 released from the effluent (kg CH_4)	–	Model value	–	BioWin
CH_4 produced during anaerobic sludge digestion (kg CH_4)	–	Model value	–	BioWin
Soluble CH_4 released from excess sludge (kg CH_4)	–	Model value	–	BioWin
N_2O released during biological treatment (kg N_2O-N/kg N)	0.003	0.035	0.253	Foley et al. (2010a, 2010b)
N_2O released from the effluent (kg N_2O-N/kg N)	0.0005	0.0025	0.005	IPCC (1997, 2006a)
N_2O directly released from excess sludge (kg N_2O-N/kg N)	0.003	0.01	0.03	Doka (2003), IPCC (2006a)
CO_2 produced from CH_4 combustion for energy generation (g CO_2-e/kWh)	–	353	–	IPCC (2001)
CO_2 produced during electricity consumption (g CO_2-e/kWh)	–	877	–	IPCC (2001)

Note Median values were selected for case study calculations

4.5.3.4 Amount of Recovered Bioenergy

The amount of recovered bioenergy (q_{bioe}) can be determined using the following equation:

$$q_{\text{bioe}} = \frac{Q_{\text{CH}_4} \times H_{\text{CH}_4} \times h}{Q}, \tag{4.7}$$

where q_{bioe} is the amount of recovered bioenergy (kWh/m^3); Q_{CH_4} is the amount of CH_4 generated during the anaerobic digestion of sludge (m^3/d), which can be obtained from the BioWin simulation; H_{CH_4} is the heating value of CH_4 (kWh/m^3), which was set to 11.67 kWh/m^3; and η is the energy conversion efficiency of CH_4 combustion, which was set to 75%.

4.5.3.5 Amount of Recovered Nutrients

The amount of recovered nutrients, including nitrogen (q_{nitr}, kg N/m^3) and phosphorus (q_{phos}, kg P/m^3), was directly obtained from the BioWin simulation.

4.5.3.6 Amount of Recovered Struvite

The amount of recovered struvite (q_{stru}, kg ISS/m^3) was directly obtained from the BioWin simulation.

4.5.4 Normalization of the Assessment Indicators

4.5.4.1 Definition of the Normalization

The assessment indicator system plays a nonnegligible role in the assessment system of multi-objective management of municipal wastewater treatment. However, due to differences in the units and numerical scales among various quantitative indicators, direct mathematical comparisons of the indicators are not possible. To avoid incommensurability among the assessment indicators, the values of these indicators must be normalized. In this subsection, a simple and rapid data processing algorithm was developed for the rapid normalization of the indicator values. Firstly, the maximum and minimum values of each indicator in the indicator system were identified; subsequently, the maxima and minima were used to normalize the corresponding datasets, as shown in Eqs. (4.8) and (4.9):

$$\text{NF}_i = \frac{p_i - \min(p_1, p_2, \ldots, p_n)}{\max(p_1, p_2, \ldots, p_n) - \min(p_1, p_2, \ldots, p_n)} \tag{4.8}$$

$$\text{PF}_i = \frac{q_i - \min(q_1, q_2, \ldots, q_n)}{\max(q_1, q_2, \ldots, q_n) - \min(q_1, q_2, \ldots, q_n)}, \tag{4.9}$$

where NF is a negative factor (dimensionless) with values ranging between 0 and 1, which represents assessment indicators such as energy consumption, chemical consumption, and GHG emissions; PF is a positive factor (dimensionless) with values ranging between 0 and 1, which represents assessment indicators such as bioenergy recovery, nutrient recovery, and struvite recovery; p represents the raw data of inventory items such as energy consumption, chemical consumption, and GHG emissions;

and q represents the raw data of inventory items such as bioenergy recovery, nutrient recovery, and struvite recovery.

An NF or PF value of 0 indicates that the impact of the corresponding assessment indicator is negligible, while an NF or PF value of 1 indicates that the impact of the corresponding assessment indicator is the most significant.

4.5.4.2 Negative Factors

Because the energy consumption, chemical consumption, and GHG emissions have negative impacts on the environment, these three indicators were defined as negative factors that negatively contribute to the overall benefits. The respective normalization formulae are as follows:

$$\mathrm{NF}_{\mathrm{chem},i} = \frac{p_{\mathrm{chem},i} - \min(p_{\mathrm{chem},1}, p_{\mathrm{chem},2}, \ldots, p_{\mathrm{chem},n})}{\max(p_{\mathrm{chem},1}, p_{\mathrm{chem},2}, \ldots, p_{\mathrm{chem},n}) - \min(p_{\mathrm{chem},1}, p_{\mathrm{chem},2}, \ldots, p_{\mathrm{chem},n})} \tag{4.10}$$

$$\mathrm{NF}_{\mathrm{ener},\,i} = \frac{p_{\mathrm{ener},i} - \min(p_{\mathrm{ener},1}, p_{\mathrm{ener},2}, \ldots, p_{\mathrm{ener},n})}{\max(p_{\mathrm{ener},1}, p_{\mathrm{ener},2}, \ldots, p_{\mathrm{ener},\,\mathrm{n}}) - \min(p_{\mathrm{ener},1}, p_{\mathrm{ener},2}, \ldots, p_{\mathrm{ener},n})} \tag{4.11}$$

$$\mathrm{NF}_{\mathrm{gree},i} = \frac{p_{\mathrm{gree},i} - \min(p_{\mathrm{gree},1}, p_{\mathrm{gree},2}, \ldots, p_{\mathrm{gree},n})}{\max(p_{\mathrm{gree},1}, p_{\mathrm{gree},2}, \ldots, p_{\mathrm{gree},n}) - \min(p_{\mathrm{gree},1}, p_{\mathrm{gree},2}, \ldots, p_{\mathrm{gree},n})}, \tag{4.12}$$

where $\mathrm{NF}_{\mathrm{chem}}$ is the normalized chemical consumption indicator (dimensionless); $\mathrm{NF}_{\mathrm{ener}}$ is the normalized energy consumption indicator (dimensionless); $\mathrm{NF}_{\mathrm{gree}}$ is the normalized GHG emissions indicator (dimensionless); $p_{\mathrm{chem},i}$ is the chemical consumption datum of the ith alternative wastewater treatment process (kg/m^3); $p_{\mathrm{ener},i}$ is the energy consumption datum of the ith alternative wastewater treatment process (kWh/m^3); and $p_{\mathrm{gree},i}$ is the GHG emissions datum of the ith alternative wastewater treatment process (kg CO_2-e/m^3).

4.5.4.3 Positive Factors

Considering that the recovery of bioenergy, nutrients, and struvite have positive environmental benefits, these indicators were defined as positive factors. The respective normalization formulae are as follows:

$$\mathrm{PF}_{\mathrm{bioe},i} = \frac{q_{\mathrm{bioe},i} - \min(q_{\mathrm{bioe},1}, q_{\mathrm{bioe},2}, \ldots, q_{\mathrm{bioe},n})}{\max(q_{\mathrm{bioe},1}, q_{\mathrm{bioe},2}, \ldots, q_{\mathrm{bioe},n}) - \min(q_{\mathrm{bioe},1}, q_{\mathrm{bioe},2}, \ldots, q_{\mathrm{bioe},n})} \tag{4.13}$$

$$\mathrm{PF}_{\mathrm{nitr},i} = \frac{q_{\mathrm{nitr,i}} - \min(q_{\mathrm{nitr},1}, q_{\mathrm{nitr},2}, \ldots, q_{\mathrm{nitr},n})}{\max(q_{\mathrm{nitr},1}, q_{\mathrm{nitr},2}, \ldots, q_{\mathrm{nitr},\,n}) - \min(q_{\mathrm{nitr},1}, q_{\mathrm{nitr},2}, \ldots, q_{\mathrm{nitr},n})} \tag{4.14}$$

$$\mathrm{PF}_{\mathrm{phos},i} = \frac{q_{\mathrm{phos},i} - \min(q_{\mathrm{phos},1}, q_{\mathrm{phos},2}, \ldots, q_{\mathrm{phos},n})}{\max(q_{\mathrm{phos},1}, q_{\mathrm{phos},2}, \ldots, q_{\mathrm{phos},n}) - \min(q_{\mathrm{phos},1}, q_{\mathrm{phos},2}, \ldots, q_{\mathrm{phos},n})} \tag{4.15}$$

$$\mathrm{PF}_{\mathrm{stru},i} = \frac{q_{\mathrm{stru},\,i} - \min(q_{\mathrm{stru},1}, q_{\mathrm{stru},2}, \ldots, q_{\mathrm{stru},n})}{\max(q_{\mathrm{stru},1}, q_{\mathrm{stru},2}, \ldots, q_{\mathrm{stru},n}) - \min(q_{\mathrm{stru},1}, q_{\mathrm{stru},2}, \ldots, q_{\mathrm{stru},n})}, \tag{4.16}$$

where PF_{bioe} is the normalized energy recovery indicator (dimensionless); PF_{nitr} is the normalized nitrogen fertilizer recovery indicator (dimensionless); PF_{phos} is the normalized phosphorus fertilizer recovery indicator (dimensionless); PF_{stru} is the normalized struvite recovery indicator (dimensionless); $p_{bioe,i}$ is the energy recovery datum of the ith alternative wastewater treatment process (kWh/m^3); $p_{nitr,i}$ is the nitrogen recovery datum of the ith alternative wastewater treatment process (kg N/m^3); $p_{phos,i}$ is the phosphorus recovery datum of the ith alternative wastewater treatment process (kg P/m^3); and $p_{stru,i}$ is the struvite recovery datum of the ith alternative wastewater treatment process (kg ISS/m^3).

4.5.4.4 Synthesized Factors

Based on the calculated negative factors (NFs) and positive factors (PFs), synthesized factors (SFs) for the four scenarios were proposed to facilitate scenario analysis. The SFs range from −1 to 1 and represent the degree of favorableness of a certain scenario: An SF value of −1 indicates that the scenario is the most unfavorable, while an SF value of 1 indicates that the scenario is the most favorable.

The SF of the environmental footprint scenario (SF_{envi}, dimensionless) was determined using the following equation:

$$SF_{envi,i} = -\frac{SF_{orig\text{-}envi,i} - MIN_{envi}}{MAX_{envi} - MIN_{envi}}, i = 1, 2, 3, \ldots, n, \tag{4.17}$$

where $SF_{orig\text{-}envi,i}$, MIN_{envi}, and MAX_{envi} were calculated as follows:

$$SF_{orig\text{-}envi,i} = NF_{orig\text{-}envi,i} = w_{chem}NF_{chem,i} + w_{ener}NF_{ener,i} + w_{gree}NF_{gree,i} \tag{4.18}$$

$$MIN_{envi} = \min\left(SF_{orig\text{-}envi,1}, SF_{orig\text{-}envi,2}, \ldots, SF_{orig\text{-}envi,n}\right) \tag{4.19}$$

$$MAX_{envi} = \max\left(SF_{orig\text{-}envi,1}, SF_{orig\text{-}envi,2}, \ldots, SF_{orig\text{-}envi,n}\right), \tag{4.20}$$

where w_{ener}, w_{chem}, and w_{gree} are the weight factors of the energy consumption, chemical consumption, and GHG emissions.

The SF of the bioenergy recovery scenario (SF_{bioe}, dimensionless) was determined using the following equation:

$$SF_{bioe,i} = PF_{norm\text{-}bioe,i} - NF_{norm\text{-}bioe,i}, i = 1, 2, 3, \ldots, n, \tag{4.21}$$

where $PF_{norm\text{-}bioe,i}$ and $NF_{norm\text{-}bioe,i}$ were calculated as follows:

$$PF_{norm\text{-}bioe,i} = \frac{PF_{orig\text{-}bioe,i} - MIN_{bioe,PF}}{MAX_{bioe,PF} - MIN_{bioe,PF}} \tag{4.22}$$

$$\mathrm{NF}_{\text{norm-bioe},i} = \frac{\mathrm{NF}_{\text{orig-bioe},i} - \mathrm{MIN}_{\text{bioe,NF}}}{\mathrm{MAX}_{\text{bioe,NF}} - \mathrm{MIN}_{\text{bioe,NF}}}, \tag{4.23}$$

and $\mathrm{PF}_{\text{orig-bioe},i}$, $\mathrm{NF}_{\text{orig-bioe},i}$, $\mathrm{MIN}_{\text{bioe,PF}}$, $\mathrm{MAX}_{\text{bioe,PF}}$, $\mathrm{MIN}_{\text{bioe,NF}}$, and $\mathrm{MAX}_{\text{bioe,NF}}$ were obtained using the following equations:

$$\mathrm{PF}_{\text{orig-bioe},i} = w_{\text{bioe}}\mathrm{PF}_{\text{bioe},i} \tag{4.24}$$

$$\mathrm{MIN}_{\text{bioe,PF}} = \min\left(\mathrm{PF}_{\text{orig-bioe},1}, \mathrm{PF}_{\text{orig-bioe},2}, \ldots, \mathrm{PF}_{\text{orig-bioe},n}\right) \tag{4.25}$$

$$\mathrm{MAX}_{\text{bioe,PF}} = \max\left(\mathrm{PF}_{\text{orig-bioe},1}, \mathrm{PF}_{\text{orig-bioe},2}, \ldots, \mathrm{PF}_{\text{orig-bioe},n}\right) \tag{4.26}$$

$$\mathrm{NF}_{\text{orig-bioe},i} = w_{\text{chem}}\mathrm{NF}_{\text{chem},i} + w_{\text{ener}}\mathrm{NF}_{\text{ener},i} + w_{\text{gree}}\mathrm{NF}_{\text{gree},i} \tag{4.27}$$

$$\mathrm{MIN}_{\text{bioe,NF}} = \min\left(\mathrm{NF}_{\text{orig-bioe},1}, \mathrm{NF}_{\text{orig-bioe},2}, \ldots, \mathrm{NF}_{\text{orig-bioe},n}\right) \tag{4.28}$$

$$\mathrm{MAX}_{\text{bioe,NF}} = \max\left(\mathrm{NF}_{\text{orig-bioe},1}, \mathrm{NF}_{\text{orig-bioe},2}, \ldots, \mathrm{NF}_{\text{orig-bioe},n}\right), \tag{4.29}$$

where w_{bioe} is the weight factor of bioenergy recovery.

The SF of the nutrient recovery scenario ($\mathrm{SF}_{\text{bioe}}$, dimensionless) was determined using the following equation:

$$\mathrm{SF}_{\text{nutr},i} = \mathrm{PF}_{\text{norm-nutr},i} - \mathrm{NF}_{\text{norm-nutr},i},\ i = 1, 2, 3, \ldots, n, \tag{4.30}$$

where $\mathrm{PF}_{\text{norm-nutr},i}$ and $\mathrm{NF}_{\text{norm-nutr},i}$ were calculated as follows:

$$\mathrm{PF}_{\text{norm-nutr},i} = \frac{\mathrm{PF}_{\text{orig-nutr},i} - \mathrm{MIN}_{\text{nutr,PF}}}{\mathrm{MAX}_{\text{nutr,PF}} - \mathrm{MIN}_{\text{nutr,PF}}} \tag{4.31}$$

$$\mathrm{NF}_{\text{norm-nutr},i} = \frac{\mathrm{NF}_{\text{orig-nutr},i} - \mathrm{MIN}_{\text{nutr,NF}}}{\mathrm{MAX}_{\text{nutr,NF}} - \mathrm{MIN}_{\text{nutr,NF}}}, \tag{4.32}$$

and $\mathrm{PF}_{\text{orig-nutr},i}$, $\mathrm{NF}_{\text{orig-nutr},i}$, $\mathrm{MIN}_{\text{nutr,PF}}$, $\mathrm{MAX}_{\text{nutr,PF}}$, $\mathrm{MIN}_{\text{nutr,NF}}$, and $\mathrm{MAX}_{\text{nutr,NF}}$ were obtained using the following equations:

$$\mathrm{PF}_{\text{orig-nutr},i} = w_{\text{nitr}}\mathrm{PF}_{\text{nitr},i} + w_{\text{phos}}\mathrm{PF}_{\text{phos},i} \tag{4.33}$$

$$\mathrm{MIN}_{\text{nutr,PF}} = \min\left(\mathrm{PF}_{\text{orig-nutr},1}, \mathrm{PF}_{\text{orig-nutr},2}, \ldots, \mathrm{PF}_{\text{orig-nutr},n}\right) \tag{4.34}$$

$$\mathrm{MAX}_{\text{nutr,PF}} = \max\left(\mathrm{PF}_{\text{orig-nutr},1}, \mathrm{PF}_{\text{orig-nutr},2}, \ldots, \mathrm{PF}_{\text{orig-nutr},n}\right) \tag{4.35}$$

$$\mathrm{NF}_{\text{orig-nutr},\,i} = w_{\text{chem}}\mathrm{NF}_{\text{chem},i} + w_{\text{ener}}\mathrm{NF}_{\text{ener},i} + w_{\text{gree}}\mathrm{NF}_{\text{gree},i} \tag{4.36}$$

$$\mathrm{MIN}_{\text{nutr,nf}} = \min\left(\mathrm{NF}_{\text{orig-nutr},1}, \mathrm{NF}_{\text{orig-nutr},2}, \ldots, \mathrm{NF}_{\text{orig-nutr},n}\right) \tag{4.37}$$

$$MAX_{nutr,nf} = \max\left(NF_{orig\text{-}nutr,1}, NF_{orig\text{-}nutr,2}, \ldots, NF_{orig\text{-}nutr,n}\right), \quad (4.38)$$

where w_{nitr} and w_{phos} are the weight factors of the nutrient recovery.

The SF of the simultaneous bioenergy and nutrient recovery scenario (SF_{simu}, dimensionless) was determined with the following equation:

$$SF_{simu,i} = PF_{norm\text{-}simu,i} - NF_{norm\text{-}simu,i}, i = 1, 2, 3 \ldots, n, \quad (4.39)$$

where $PF_{norm\text{-}simu,i}$ and $NF_{norm\text{-}simu,i}$ were calculated as follows:

$$PF_{norm\text{-}simu,i} = \frac{PF_{orig\text{-}simu,i} - MIN_{simu,PF}}{MAX_{simu,PF} - MIN_{simu,PF}} \quad (4.40)$$

$$NF_{norm\text{-}simu,i} = \frac{NF_{orig\text{-}simu,i} - MIN_{simu,NF}}{MAX_{simu,NF} - MIN_{simu,NF}}, \quad (4.41)$$

and $PF_{orig\text{-}simu,i}$, $NF_{orig\text{-}simu,i}$, $MIN_{simu,PF}$, $MAX_{simu,PF}$, $MIN_{simu,NF}$, and $MAX_{simu,NF}$ were obtained using the following equations:

$$PF_{orig\text{-}simu,i} = w_{bioe}PF_{bioe,i} + w_{nitr}PF_{nitr,i} + w_{phos}PF_{phos,i} + w_{stru}PF_{stru,i} \quad (4.42)$$

$$MIN_{simu,PF} = \min\left(PF_{orig\text{-}simu,1}, PF_{orig\text{-}simu,2}, \ldots, PF_{orig\text{-}simu,n}\right) \quad (4.43)$$

$$MAX_{simu,PF} = \max\left(PF_{orig\text{-}simu,1}, PF_{orig\text{-}simu,2}, \ldots, PF_{orig\text{-}simu,n}\right) \quad (4.44)$$

$$NF_{orig\text{-}simu,i} = w_{chem}NF_{chem,i} + w_{ener}NF_{ener,i} + w_{gree}NF_{gree,i} \quad (4.45)$$

$$MIN_{simu,NF} = \min\left(NF_{orig\text{-}simu,1}, NF_{orig\text{-}simu,2}, \ldots, NF_{orig\text{-}simu,n}\right) \quad (4.46)$$

$$MAX_{simu,NF} = \max\left(NF_{orig\text{-}simu,1}, NF_{orig\text{-}simu,2}, \ldots, NF_{orig\text{-}simu,n}\right), \quad (4.47)$$

where w_{stru} is the weight factor of the struvite recovery.

4.6 Data Quality Analysis for the Assessment of Multi-objective Management

4.6.1 Analysis Methods

Data quality reflects the degree to which data satisfy specific applications. For multi-objective management systems, basic data and the quality of such data play a critical role in determining the correctness of the assessment. Therefore, based on the sources of uncertainty of basic data, data uncertainty analysis was performed from six aspects (Lindfors 1995): (1) reliability of the data source; (2) independency of samples; (3) credibility of data; (4) representativeness of the year; (5) representativeness of the geographical location; and (6) representativeness of the technology. These six assessment indicators encompass the main factors that influence the quality of basic data. To assess the data quality level, points were awarded based on a five-point scale, where one point represents high-quality data and five points represent low-quality data (Table 4.2).

4.6.2 Analysis Results

Based on the scoring criteria described above, the quality of the basic data used in the present study was assessed. Based on the results shown in Table 4.3, the data uncertainty is mainly manifested in the estimation of the N_2O emissions. In the present study, it was assumed that N_2O emissions primarily originate from the wastewater treatment process, fugitive emissions from the effluent, and volatilization from sludge. Because there is a lack of dynamic predictive models for the N_2O release during wastewater treatment, emission factors from the previous literature were used for the estimation of the N_2O emissions. Therefore, a certain degree of uncertainty exists in the raw data. With regard to the energy consumption of other parts of the WWTP (e.g., lighting), the estimation was performed based on limited statistical data because investigations of this aspect are relatively rare, thereby leading to a certain degree of uncertainty. In general, the sources of basic data required for the assessment framework have a high degree of controllability and the data were of high numerical quality.

4.7 Summary

Based on the assurance of the water quality, a concept of multi-objective management of municipal wastewater treatment aiming at energy conservation, reduced carbon emissions, and resource recovery from pollutants has been proposed. By following the principles of indicator system construction, minimization of the environmental footprint, and maximization of resource recovery, an assessment system for the multi-objective management of municipal wastewater treatment was constructed from several aspects such as boundary definition, data acquisition, and data processing. Concurrently, according to the requirements for the sustained management of wastewater treatment and considering the technological levels of existing wastewater and sludge treatment and disposal methods, four scenarios were con-

Table 4.2 Matrix used for data quality assessment

Area assessed	Scoring criteria				
	1	2	3	4	5
Reliability of the data source (DQA1)	Raw data obtained from measurements	Data calculated based on measurement methods	Some data calculated based on assumptions	Data obtained based on assessment(s) by leading expert(s)	Data obtained through unreliable assessment(s)
Independency of samples (DQA2)	Validated data samples from public or independent databases	Validated sample data from the studied organization	Independent data samples from unvalidated sources within organizations	Data from unvalidated sources within organizations	Data from unvalidated sources within the studied organization
Credibility of data (DQA3)	Adequate sample size, appropriate data collection duration	Slightly smaller data range but with appropriate data collection duration	Adequate data range but with a slightly shorter data collection duration	Small data range and short data collection duration or inadequate data range and inadequate collection duration	Data from unknown range and duration or inadequate data from a small data range and short data collection duration
Representativeness of the year (DQA4)	Time-independent data or data from the last 3 years	Data from the last 6 years	Data from the last 10 years	Data from the last 15 years	Data from unknown year(s)
Representativeness of the geographical location (DQA5)	Data from the study area	Average data from a larger area, which includes the study area	Data from areas with highly similar production conditions and productivity levels	Data from areas with moderately similar production conditions and productivity levels	Data from unknown areas or from areas with completely different production conditions and productivity levels

(continued)

Table 4.2 (continued)

Area assessed	Scoring criteria				
	1	2	3	4	5
Representativeness of the technology (DQA6)	Data obtained from the studied organization	Identical technology, processes, and raw materials but different organization	Identical technology but different processes and raw materials	Different technology but identical products	Data of similar products used due to the lack of data

Adapted from Wang et al. (2012), with permission from American Chemical Society, Copyright (2012)

Table 4.3 Results of the data quality assessment

Data type	DQA1	DQA2	DQA3	DQA4	DQA5	DQA6	Average
Chemical consumption	1	3	1	2	2	1	1.7
Energy consumption of aeration	1	3	1	2	2	1	1.7
Energy consumption of the water pump	2	3	1	2	2	1	1.8
Energy consumption of the mixer	2	3	1	2	2	1	1.8
Energy consumption of heat loss	2	3	1	2	2	1	1.8
Other energy consumption	3	4	2	3	3	2	2.8
CH_4	2	3	1	2	2	1	1.8
N_2O	3	1	3	3	3	3	2.7
CO_2	2	3	1	2	2	1	1.8
Energy recovered	2	3	1	2	2	1	1.8
Nutrients recovered	1	3	1	2	2	1	1.7
Struvite recovered	1	3	1	2	2	1	1.7

Adapted from Wang et al. (2012), with permission from American Chemical Society, Copyright (2012)

structed for the first time for scenario analysis. Lastly, the basic data involved in the assessment system and the sources of such data were analyzed and the analysis results indicate that the data acquisition methods and data sources established in the assessment framework are highly reliable.

References

Doka, G. (2003). Life cycle inventories of waste treatment services (Ecoinvent Report No. 13), Swiss Centre for Life Inventories, Dubendorf.

Foley, J., de Haas, D., Hartley, K., & Lant, P. (2010a). Comprehensive life cycle inventories of alternative wastewater treatment systems. *Water Research, 44*(5), 1654–1666.

Foley, J., de Haas, D., Yuan, Z. G., & Lant, P. (2010b). Nitrous oxide generation in full-scale biological nutrient removal wastewater treatment plants. *Water Research, 44*(3), 831–844.

Horne, R., Grant, T., & Verghese, K. (2009). *Life cycle assessment: Principles, practice and prospects*. Victoria, Australia: CSIRO Publishing.

IPCC. (1997). Reference manual: Intergovernmental panel on climate change. Geneva: the National Greenhouse Gas Inventories Programme.

IPCC. (2001). *Climate change 2001: The scientific basis*. Cambridge: Cambridge University Press.

IPCC. (2006a). IPCC guidelines for national greenhouse gas inventories, National Greenhouse Gas Inventories Programme.

IPCC. (2006b). Wastewater treatment and discharge. H.S. Eggleston, L. Buendia, K. Miwa, T. Ngara, & K. Tanabe (Eds.), The National Greenhouse Gas Inventories Programme, Japan.

Le Corre, K. S., Valsami-Jones, E., Hobbs, P., & Parsons, S. A. (2009). Phosphorus recovery from wastewater by struvite crystallization: A review. *Critical Reviews in Environmental Science and Technology, 39*(6), 433–477.

Lindfors, L.-G. (1995). *Nordic guidelines on life-cycle assessment*. Copenhagen: Nordic Council of Ministers.

Logan, B. E., Hamelers, B., Rozendal, R. A., Schrorder, U., Keller, J., Freguia, S., et al. (2006). Microbial fuel cells: Methodology and technology. *Environmental Science and Technology, 40*(17), 5181–5192.

Metcalf, I., & Eddy, H. (2003). *Wastewater engineering: Treatment and reuse*. New York: McGraw-Hill.

Ren, N. Q., Guo, W. Q., Liu, B. F., Cao, G. L., & Ding, J. (2011). Biological hydrogen production by dark fermentation: Challenges and prospects towards scaled-up production. *Current Opinion in Biotechnology, 22*(3), 365–370.

Shahabadi, M. B., Yerushalmi, L., & Haghighat, F. (2009). Impact of process design on greenhouse gas (GHG) generation by wastewater treatment plants. *Water Research, 43*(10), 2679–2687.

Shaw, A., Kadava, A., & Tarallo, S. (2011). *Refinement of life cycle assessment (LCA) Methods for water and wastewater treatment plant design*. Amsterdam: IWA publisher.

Verstraete, W., & Vlaeminck, S. E. (2011). ZeroWasteWater: Short-cycling of wastewater resources for sustainable cities of the future. *International Journal of Sustainable Development and World Ecology, 18*(3), 253–264.

Wang, X., Liu, J. X., Ren, N. Q., Yu, H. Q., Lee, D. J., & Guo, X. S. (2012). Assessment of Multiple sustainability demands for wastewater treatment alternatives: A Refined evaluation scheme and case study. *Environmental Science and Technology, 46*(10), 5542–5549.

References

Chapter 5
Determination of the Weighting Element of Assessment Indicators

5.1 Overview

The assessment system for multi-objective management of wastewater treatment is a comprehensive system that involves multiple scenarios and objectives. Since the various indicators in the indicator system have different roles and levels of importance, different weighting coefficients must be assigned to the different assessment indicators. Indicator weights, which reflect the different levels of importance of the indicators in an assessment system, are comprehensive measures of the subjective assessment and objective reflections of the relative importance of indicators in an assessment or decision-making process. The determination of the weighting system and the reasonableness of the assigned weights are critical in determining the scientific reasonableness of the assessment conclusions. A change in the weight of a certain indicator will influence all assessment conclusions. In addition, under different spatial, temporal, and geographical conditions, the level of importance of the same assessment indicator for the same assessed subject may be different; therefore, the weights of the respective assessment indicators must be determined based on actual circumstances.

In this chapter, global time series databases were used to define a weighting system for the assessment system for multi-objective management based on the construction of an assessment system for the multi-objective management of municipal wastewater treatment discussed in Chap. 4. Statistical principles were employed to test the irreplaceability of various indicators of the indicator system constructed in Chap. 4 to further optimize the indicator system. Subsequently, weight prediction models based on historical data were constructed in preparation for the scenario analysis for multi-objective management of municipal wastewater treatment in Chap. 6.

X. Wang, *Energy Consumption, Chemical Use and Carbon Footprints of Wastewater Treatment Alternatives*, Springer Theses,
https://doi.org/10.1007/978-981-13-5983-5_5

5.2 Indicator Weight Characteristics

For multi-indicator assessments, predictions, and multi-objective decision-making processes, weights, also known as weighting coefficients, are used to represent the relative importance of the indicators and the level of importance of each indicator for the assessed subject. Generally speaking, weights have the following main characteristics: (1) weights reflect the objective and value of the assessment and highlight the key points within certain periods or areas that require attention; (2) weights affect the assessment conclusions to a great extent because the weights of various assessment indicators are mutually constrained; when the weight of an indicator is set to a larger value, the weights of other indicators will be smaller and changes in the weights may also alter the assessment outcomes and result in changed decisions; and (3) when the assessment intent changes, the weighting coefficient of the same assessment indicator will have to be changed. In an assessment system, weights should be appropriately determined and adjusted based on the relative importance of the assessment intents and indicators, i.e., the effects of factors reflected by assessment indicators on the comprehensive assessment. Weights should reflect the respective importance of each factor for the comprehensive assessment to enhance the comparability of the assessment indicators.

5.3 Methods for Weight Determination

Weights inherently contain important information about the comprehensive assessment, which influences the feasibility of the assessment conclusions. Therefore, weights should be widely acknowledged and relatively stable. Assessment indicators must be continuously re-understood, re-examined, and updated, and the corresponding weights must be continuously revised. When determining the indicator weights, the following principles should be followed: (1) objectivity, (2) rangeability, (3) relevance, (4) comparability, and (5) hierarchicality. Based on the reasonable determination of the weighting system, the relative levels of importance of various indicators for the assessed subject must be distinguished; at the same time, the distortion of the assessment conclusions must also be prevented.

Indicator weights can be determined with many methods, which can be broadly classified into two categories: (1) Objective weight assignment, where the raw data are obtained through calculations using actual data of the indicators in assessment units, for example, the mean squared error method, maximizing deviation method, simple correlation function method, and entropy weight method; (2) Subjective weight assignment, where the raw data are determined by experts based on experience, for example, the analytic hierarchy process and expert investigation method. These two categories of weighting methods have their own advantages and disadvantages: Objective weight assignment methods produce weights with a higher accuracy but lower explainability; that is, it is difficult to explain the obtained results. Subjective weight assignment methods lead to less objectivity but higher explainability (refer to Sect. 2.5). During the determination of the weighting system, the coexistence of various indicators should be appropriately handled, weaknesses of different weight assignment methods should be overcome, and objective weights obtained through data mining and subjective selections formed through experience and knowledge-based assessments should be harmoniously combined to the greatest extent possible to obtain the final weighting system. Therefore, the strengths of the objective and subjective weight assignment methods were combined in the present study, and the perspective of subjective experience was adopted to analyze the problems and possibly related data sources fed back by various assessment indicators. In addition, objective historical databases were employed for the mining of characteristic data of different countries from a global standpoint, which was used for the construction of the weighting system.

5.4 Steps for the Determination of Indicator Weights

Weighting based on historical data, which is an indicator weight determination method with a combined qualitative and quantitative approach, is proposed in the present study. It is applicable to decision-making problems with complex structures, multiple decision criteria, and importance levels that are difficult to quantify. In addition, it enables the harmonious combination of subjective judgments and deductions of decision-makers, thereby leading to systemization, modelization, and digitalization of the decision-making process. The basic steps of weighting based on historical data are as follows: Firstly, the weights of the various assessment indicators in the assessment system for multi-objective management were defined based on experience and the main data that objectively reflect the importance of the indicators in the time series databases were identified. Next, historical data mining was performed based on the defined weights of various indicators and the absolute weight matrices of various countries were obtained through data processing. Subsequently, statistical principles were applied for the analysis of correlations among indicator weights to test the irreplaceability of various indicators in the indicator system and to construct weight prediction models based on historical data. The detailed steps are described in the following subsections.

5.4.1 Definition of Weights

The weights of the indicators were defined based on the problems fed back by various indicators in the assessment system of multi-objective management. The main data that objectively reflect the importance of the indicators in the time series database were identified to determine the data sources of the indicators. The weighting equation for various indicators is:

$$\omega_j(x) = \frac{S_{\text{obj},j}}{S_{\text{ref},j}}, \tag{5.1}$$

where ω_j (x) is the absolute weight of assessment indicator j in the target year (x); $S_{\text{obj},j}$ is the value of assessment indicator j in the target year (x); and $S_{\text{ref},j}$ is the value of assessment indicator j in the reference year.

5.4.2 Selection of the Reference Year

Data from the reference year form the basis of the scenario design and policy analysis. The concept of the reference year is widely applied in many fields including banking, construction, water resources, and climate change. At present, it is most commonly used in climate change mitigation (e.g., reduction of greenhouse gas (GHG) emissions and GHG inventories). For instance, 1990 was set as the reference year for emissions reduction in the Kyoto Protocol. The reference year can be any year with quantitative data or several years from which historical average data can be obtained. Except for the GHG inventories, specific reference years have not been set for the aspects of energy consumption, chemical consumption, and resource recovery. Therefore, 1990, which is the reference year of the Kyoto Protocol, was also used as the reference year in the present study.

5.4.3 Data Mining

To obtain basic data for the quantification of indicator weights, relevant global time series databases were integrated based on the weight definitions of various indicators and the time period of 1990–2010 was selected for data mining. Specifically, data from 1990 were used as reference year data, data from 1991 to 1999 were used for the construction of the weight prediction model, and data from 2010 were used for the validation of the weight prediction model. The databases used in the integration of the data sources include the database provided by the US Energy Information Administration, Library of Economic Cooperation and Development, and the British Geological Survey Database supported by the Natural Environment Research Council.

5.4.4 Selection of Countries and Regions

The present study mainly involves comparative studies of developed and developing countries. In view of the data coverage in the time series databases, the following 26 developed countries were considered Canada, USA, Austria, Belgium, Denmark , Finland, France, Greece, Hungary, Iceland, Ireland, Italy, Luxembourg,

Netherlands, Norway, Poland, Portugal, Spain, Sweden, Switzerland, Turkey, UK, Israel, Australia, Japan, and South Korea. The following 26 developing countries were considered Mexico, Argentina, Barbados, Brazil, Chile, Colombia, Saint Lucia, Iran, Iraq, Oman, Yemen, Algeria, Benin, Cameroon, Egypt, Gabon, Mali, Morocco, South Africa, Zambia, Zimbabwe, China, Fiji, India, Malaysia, and Thailand.

5.4.5 Construction of the Weight Matrices

Based on database integration, equation in (5.1) was used to obtain the absolute weight matrix P for various countries after the basic data of the various countries that could possibly reflect the importance of indicators had been mined. The parameter U represents the assessment indicator system, ω_i represents the ith assessment indicator, $\omega_i \in (I = \text{ener, gree, chem, nutr, bioe, stru})$, and $\omega_{i,j}$ represents the relative importance of ω_i in year j (j = 1990, 1991, …, 2009).

Based on the symbols defined above, the absolute weight matrix P can be obtained using the following equation:

$$P = \begin{bmatrix} \omega_{\text{ener},1990} & \omega_{\text{ener},1991} & \cdots & \omega_{\text{ener},2009} \\ \omega_{\text{gree},1990} & \omega_{\text{gree},1991} & \cdots & \omega_{\text{gree},2009} \\ \vdots & \vdots & \vdots & \vdots \\ \omega_{\text{stru},1990} & \omega_{\text{stru},1991} & \cdots & \omega_{\text{stru},2009} \end{bmatrix}, \tag{5.2}$$

where ω_{chem} is the absolute weight of the chemical consumption indicator (dimensionless); ω_{ener} is the absolute weight of the energy consumption indicator (dimensionless); ω_{gree} is the absolute weight of the GHG emissions indicator (dimensionless); ω_{bioe} is the absolute weight of the bioenergy recovery indicator (dimensionless); ω_{nutr} is the absolute weight of the nutrient recovery indicator (dimensionless); and ω_{stru} is the absolute weight of the struvite recovery indicator (dimensionless).

5.4.6 Indicator Irreplaceability Test

In Chap. 4, six major assessment indicators (energy consumption, chemical consumption, GHG emissions, bioenergy recovery, nutrient recovery, and struvite recovery) were, respectively, selected based on different aspects of the construction of an indicator system for the assessment system of multi-objective management. During the preliminary selection of these indicators, the necessity of the indicators was primarily expounded from a literature research perspective. However, irreplaceability testing was not performed. Therefore, regression analysis was employed as statistical method for the pairwise correlation analysis of the indicator importance to determine the irreplaceability of various indicators.

5.4.7 Construction and Validation of the Weight Prediction Model

After obtaining the absolute weights of various assessment indicators, the weights were arranged in chronological order. On this basis, data analysis and mining were performed for the construction of time series-based weight prediction models. Subsequently, the accuracy of the constructed models was validated.

5.5 Determination of Indicator Weights

5.5.1 Energy Consumption

Based on baseline forecasts by the US Energy Information Administration (US EIA) (IEA 2004), the total world energy consumption will increase from 10.2 billion tons of oil equivalent (toe) in 2001 to 16.2 billion toe in 2025; that is, the total world energy consumption will increase by 54% from 2001 to 2025. Furthermore, the depletion of fossil fuels around the world is inevitable; reserves are expected to run out within the current century. Relevant data (BP 2006) indicated that the world's proven oil reserves are sufficient to meet more than 40 years of global production, while respective statistics for natural gas and coal stipulate 65 and 155 years. These data imply that higher energy consumption levels lead to higher worldwide fossil fuel depletion rates. Because the energy consumption per capita in various countries can reflect the importance of energy consumption in each country, the 1990–2009 data on energy consumption per capita of the selected countries were used as basic data to determine the energy consumption weights. After performing absolute weight determination based on the process described in Sect.5.4, the absolute weights of energy consumption for developed countries (Tables A.1 and A.2) and developing countries (Tables A.3 and A.4) were obtained. Due to space constraints, the data are not presented in weight matrices.

5.5.2 GHG Emissions

According to the global warming potentials (GWPs) with a 100-year horizon provided by the Intergovernmental Panel on Climate Change (IPCC), the GWPs of CH_4 and N_2O are 25 and 298 times that of CO_2, respectively. However, the amount of CO_2 produced through the combustion of fossil fuels and other routes cannot be overlooked. Statistics published by the World Bank (IEA 2004) indicated that the global CO_2 emissions in 2003 were 16% higher than that in 1990. The emissions produced by low- and middle-income countries accounted for only one-third of global emissions in 1960. However, from 1990 to 2003, the total emissions in China and India increased by 73% and 88%, respectively, while the total emissions in the USA and Japan increased by 20% and 15%, respectively. The total emissions increased by merely 3% in European Union (EU) member countries. A study report released by the United States National Academy of Sciences in late 2007 showed that the rate of increase of global CO_2 emissions between 2000 and 2004 was close to three times that in the 1990s. The collection and mining of data on global GHG emissions showed that there is a lack of complete data sources. However, some studies reported that CO_2 generated from energy consumption accounts for 79.5% of the total GHG emissions (standard deviation: 8.8%) and the proportion of CO_2 in GHGs is relatively stable (IEA 2004). Therefore, the 1990–2009 data on CO_2 emissions per capita (based on the energy consumption) in various countries were used as basic data to determine the absolute weights of the GHG emissions for developed countries (Tables A.5 and A.6) and developing countries (Tables A.7 and A.8).

5.5.3 Chemical Consumption

Based on a report released by the United Nations Environment Programme (Programme 2012), the global chemical sales are expected to increase by ~ 3% annually until 2050, while the chemical production between 2012 and 2020 will likely grow by an average of 40% in Africa and the Middle East and 33% in Latin America. The increased importance of the role of chemicals implies that artificially synthesized chemicals are rapidly becoming the greatest component of global pollution, thereby increasing the likelihood of humans or habitats coming into contact with harmful chemicals. Because data on the annual chemical consumption in various countries are unavailable and chemical imports by various countries are representative of the importance of chemical consumption to a certain extent, the 1990–2009 data on chemical imports by selected countries were used as basic data to determine the absolute weights of the chemical consumption for developed countries (Tables A.9 and A.10) and developing countries (Tables A.11 and A.12).

5.5.4 Bioenergy Recovery

Human progress and development have led to a continuous increase in the energy demand. To solve the problem of energy shortage and to reduce environmental pollution arising from the consumption of conventional fossil fuels, humans actively seek novel energy carriers and place greater emphasis on energy recovery. The municipal organic matter production is on an upward trend due to the continuous development of global economies and the acceleration of urbanization. Organic compounds in wastewater, excess sludge, animal waste, and other organic waste contain large amounts of energy-containing matter and energy carriers, such as CH_4 and H_2, can be generated through the conversion of such matter. Therefore, the 1999–2009 data on electricity generated from organic matter and waste in various countries were used as basic data to determine the absolute weights of bioenergy consumption in developed countries (Tables A.13 and A.14) and developing countries (Tables A.15 and A.16).

5.5.5 *Nutrient Recovery*

The excess sludge or other municipal waste generated during wastewater treatment contains substantial amounts of nutrients such as nitrogen and phosphorus. After composting, these materials can be used as soil conditioners in farmland. From another standpoint, the nutrient recovery from excess sludge or other waste material via composting provides another disposal route for organic waste, thereby reducing the environmental impacts caused by conventional landfills or incineration and solving placement issues for large amounts of sludge and waste. Therefore, during the data acquisition for the determination of weights of the nutrient recovery, the importance of nutrient recovery was not considered from the soil conditioning perspective. Instead, the generation and placement of excess sludge and other municipal waste were used to indicate the importance of the nutrient recovery in various countries or regions. In view of this perspective, the 1990–2009 data on the amount of municipal waste generated in various countries were used as basic data to determine the absolute weights of the nutrient recovery in developed countries (Tables A.17 and A.18) and developing countries (Tables A.19 and A.20).

5.5.6 *Struvite Recovery*

Phosphorus differs from nitrogen in that it mainly exists in the form of natural phosphate ores such as phosphate rock, struvite, and animal fossils. It is a nonrenewable and valuable resource because it mainly moves on a one-way path through the biosphere; the proven phosphorus reserves are expected to be exhausted in 100 years. Excess sludge discharged from wastewater treatment plants (WWTPs) contains substantial amounts of phosphorus. Several studies proved the feasibility of phosphorus recovery via struvite, which can be used as fertilizer in agriculture. Therefore, the importance of the struvite recovery was mainly considered from the perspective of phosphorus scarcity. The 1990–2009 data on phosphate rock production in various countries were used as basic data to determine the absolute weights of the struvite recovery in developed countries (Tables A.21 and A.22) and developing countries (Tables A.23 and A.24).

5.6 Irreplaceability Testing

A reasonable assessment indicator system and its constituents are extremely important for the assessment system of multi-objective management of municipal wastewater treatment. The various indicators should be combined into a system based on a

Table 5.1 R^2 values obtained from pairwise interdependency analysis of indicator weights

Assessment indicator 1	Assessment indicator 2	All countries	Developing countries	Developed countries
ω_{gree}	ω_{ener}	0.7225	0.9445[a]	0.4455
ω_{gree}	ω_{chem}	0.2095	0.0148	0.1188
ω_{gree}	ω_{bioe}	0.0353	0.0048	0.0044
ω_{gree}	ω_{nutr}	0.2458	0.1650	0.1464
ω_{gree}	ω_{stru}	0.0007	0.0232	0.1482
ω_{bioe}	ω_{chem}	0.0544	0.1081	0.0292
ω_{bioe}	ω_{ener}	0.0859	0.0043	0.0321
ω_{bioe}	ω_{nutr}	0.0298	0.1736	0.0119
ω_{bioe}	ω_{stru}	0.1317	0.1202	0.0029
ω_{stru}	ω_{ener}	0.0003	0.0314	0.3552
ω_{stru}	ω_{chem}	0.1820	0.0308	0.0001
ω_{stru}	ω_{nutr}	0.2657	0.0095	0.0156
ω_{ener}	ω_{chem}	0.1816	0.0325	0.0860
ω_{ener}	ω_{nutr}	0.2392	0.3094	0.1248
ω_{nutr}	ω_{chem}	0.1161	0.1961	0.0644

[a]Highly significant interdependency

certain hierarchical structure. In addition, they should objectively reflect the current situations and issues in the multi-objective management of wastewater treatment and form a basis for decision-making. In Sect. 4.5.2, based on the literature research, six assessment indicators were selected for the construction of an indicator system by combining the key issues of multi-objective management of wastewater treatment. However, the mutual independencies and irreplaceability of these indicators were not tested. Therefore, pairwise testing of the correlations between the assessment indicators was performed in this section using basic indicator weight data and statistical tools to determine the irreplaceability of the various indicators and optimize the assessment indicator system. If the mutual interdependency between the indicators is highly significant, the indicators have a lower irreplaceability. In this case, the indicator system can be optimized to eliminate the effects of indicator overlap; alternatively, applicability statements can be made to ensure the coordination consistency among indicators and the completeness and necessity of all indicators used. On the other hand, a weak correlation between indicators implies that the selected indicators have a higher irreplaceability. Table 5.1 shows the R^2 values obtained from pairwise comparisons of assessment weights and Fig. 5.1 shows the detailed correlation diagrams.

As shown in Table 5.1 and Fig. 5.1, the weights of the GHG emissions indicator (ω_{gree}) and energy consumption indicator (ω_{ener}) have highly significant mutual interdependencies (all countries: $R^2 = 0.7225$; developing countries: $R^2 = 0.9445$;

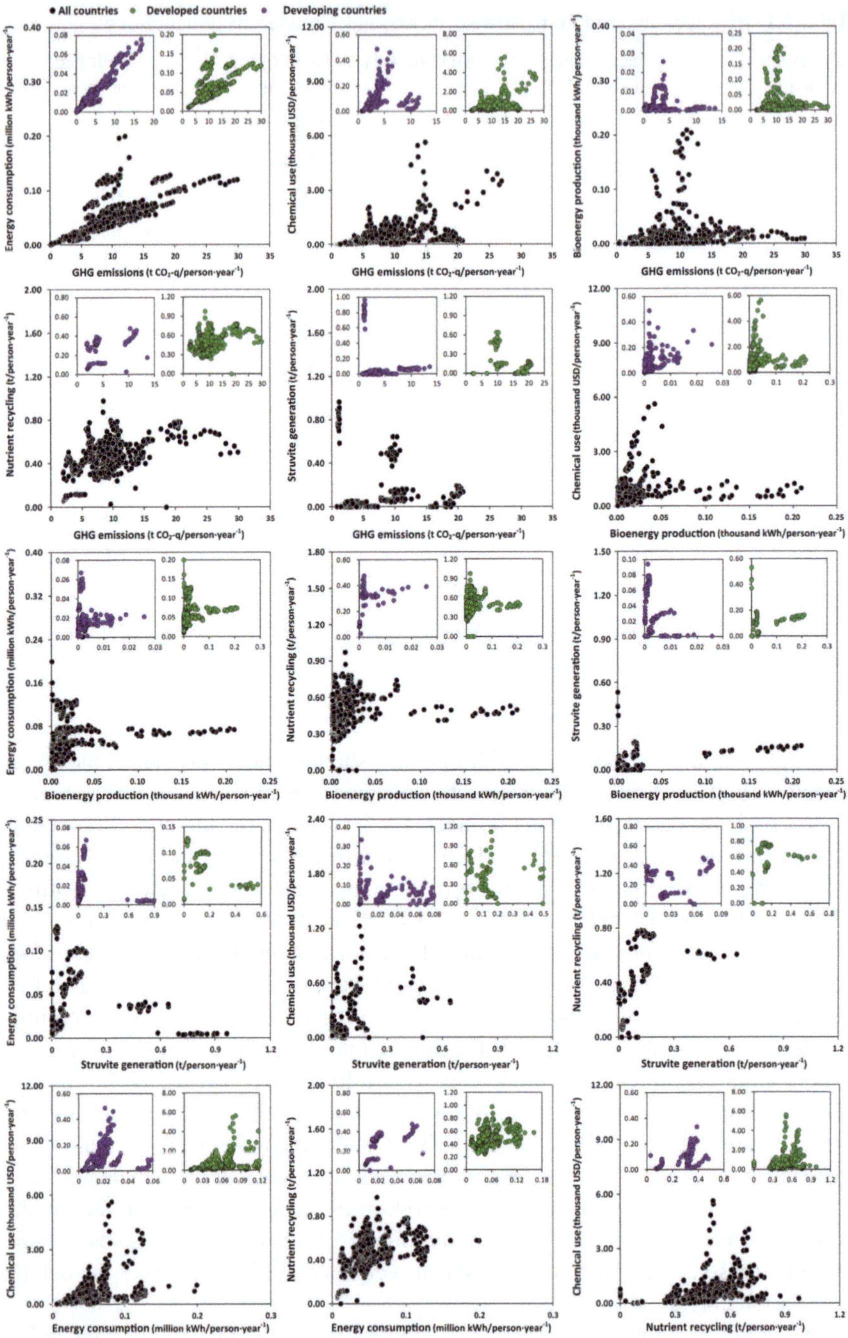

Fig. 5.1 Correlations obtained from the pairwise correlation analysis of indicator weights

developed countries: $R^2 = 0.4455$). These results show that the two indicators have a higher mutual interdependency, which suggests the possible existence of overlap between the GHG emissions indicator and energy consumption indicator; therefore, the replacement of one of the two indicators by the other indicator may be considered. Such a situation may arise because the basic data used for the quantification of the importance of GHG emissions are the CO_2 emissions generated by energy consumption (refer to the detailed assumption in Sect. 5.5.2). This indicates that the energy consumption and GHG emissions are closely related. Although a certain degree of overlap exists between the indicators, such an overlap does not merely represent repeated manifestations of the same aspect. The two indicators characterize the environmental footprint of the studied subject in two different dimensions; energy consumption mainly characterizes the consumption of fossil fuels, while GHG emissions quantify the magnitude of the climate change potential. Therefore, the energy consumption and GHG emissions indicators should coexist in the indicator system for the multi-faceted assessment of the environmental footprint of the studied subject. In general, the mutual interdependencies of any two indicators in the indicator system are not highly significant, which shows that each assessment indicator is irreplaceable to a certain extent and can be used for assessment, independent of other indicators.

5.7 Weight Prediction Models

5.7.1 Model Construction

The assessment of multi-objective management of wastewater treatment is a multi-attribute, multi-objective, and comprehensive assessment process. In the indicator system of the assessment system, various indicators have different functions, positions, and levels of importance. The weights of these indicators can reflect the importance of each indicator during the assessment process; therefore, they are comprehensive measures of the subjective assessment and objective reflection of the relative importance of indicators in an assessment or decision-making process. Because the reasonableness of the assigned weights plays a critical role in determining the scientific reasonableness of the assessment conclusions, the weight assignment process is of utmost importance. However, the determination of the indicator weights in Sect. 5.5 showed that the subjective judgment of the importance of various indicators and the acquisition and mining of objective data are both complex and time-consuming processes. Therefore, based on the quantification of historical weights of various indicators, the time series prediction method was employed for further refinement and induction of weight prediction models. Within an effective time series, the prediction models can be used to predict the weights of various indicators to forecast future development trends and provide guidance for strategic decision-making.

The time series prediction method is a prediction method based on the extrapolation of historical data and is also known as the historical extrapolation prediction method. In this method, extrapolation is performed based on the historical development and patterns of a certain phenomenon that can be reflected by time series data and extrapolated data are used to predict the future development of the phenomenon (Brockwell et al. 2002). Based on the organization and analysis of time series data, the process, direction, and trend of development of the phenomenon reflected by the time series can be used for the extrapolation or analogies to predict the levels that can be reached during the upcoming period or within a certain number of years. The method involves the following steps: (1) Acquisition and sorting of historical data associated with a certain phenomenon; (2) Identification and mining of relevant data, followed by arrangement of the data in several series; (3) Analysis of the time series to identify the patterns of the phenomenon that change with time to derive a certain mode or model; and (4) Prediction of the future status of the phenomenon using the model. Therefore, by considering the weight prediction backgrounds of developed and developing countries, the weights of the energy consumption, chemical consumption, GHG emissions, bioenergy recovery, nutrient recovery, and struvite recovery indicators within the period of 1991–2009 (1990 was not included because it was set as the reference year) were, respectively, arranged in the time series.

Subsequently, regression analysis was performed on the time series data. The general equation for the prediction models is as follows:

$$\omega(x) = a_1 x^n + a_2 x^{n-1} + a_3 x^{n-2} + \cdots + a_n x + c,\ n = 1, 2, 3, \ldots, \tag{5.3}$$

where a_1, a_2, a_3,..., a_n, and c are the unknown parameters of the model $\omega(x)$, which can be obtained from data fitting.

Based on continuous debugging, the weight prediction models with the highest significance were obtained (Tables 5.2 and 5.3).

5.7.2 *Model Validation*

The model validation process is critical to the actual application of a model. From experience, the most powerful and effective validation method involves the separation of all datasets into two portions based on the chronological order. The dataset corresponding to the first time period is used for model construction, while the dataset corresponding to the later time period is used to validate the model performance. An inability to satisfactorily validate the dataset signifies that the model has a weak extrapolation capability and thus requires further adjustments and corrections.

Table 5.2 Weight prediction models for developed countries

Indicator weight	Prediction model $\omega(x)$	R^2
ω_{gree}	$\omega_{\text{gree,ed}}(x) = -0.00121341806672573x^3 + 0.727354769181659x^2 - 1453.30424174414x + 967924.958484367$	0.92
ω_{ener}	$\omega_{\text{ener,ed}}(x) = -0.000699378007336576x^2 - 2.80983619570672x - 2820.99859045777$	0.96
ω_{chem}	$\omega_{\text{chem,ed}}(x) = 0.0000903260240093795x^3 - 0.532919420492795x^2 + 1047.90081445518x - 686730.50560836$	0.90
ω_{bioe}	$\omega_{\text{bioe,ed}}(x) = 0.000193632373924668x^3 - 1.12582261746061x^2 + 2180.20585257245x - 1406177.14233874$	0.98
ω_{nutr}	$\omega_{\text{nutr,ed}}(x) = 0.0000371406749023334x^3 - 0.223501604833942x^2 + 448.328016822373x - 299773.84054749$	0.95
ω_{stru}	$\omega_{\text{stru,ed}}(x) = 0.0000471921722318414x^5 - 0.471944976569618x^4 + 1887.86628892675x^3 - 3775892.08084515x^2 + 3776038328.27435x - 1510468530884.1$	0.95

Table 5.3 Weight prediction models for developing countries

Indicator weight	Prediction model $\omega(x)$	R^2
ω_{gree}	$\omega_{\text{gree,ing}}(x) = -0.0000288078975391459x^3 + 0.174190887880288x^2 - 351.039309526825x + 235779.400768706$	0.99
ω_{ener}	$\omega_{\text{ener,ing}}(x) = -1.30657303740178x^4 + 0.00103801642790474x^3 - 3.09103316482478x^2 + 4089.00224419342x - 2027485.20716646$	0.99
ω_{chem}	$\omega_{\text{chem,ing}}(x) = 0.018798307725162x^2 - 74.9383888441255x + 74685.7256819945$	0.90
ω_{bioe}	$\omega_{\text{bioe,ing}}(x) = 0.00732847870222575x^2 - 28.9932867119523x + 28675.9111265406$	0.98
ω_{nutr}	$\omega_{\text{nutr,ing}}(x) = 0.0000121414792255414x^4 - 0.0971135587990962x^3 + 291.283319727665x^2 - 388297.846855507x + 194107219.308722$	0.99
ω_{stru}	$\omega_{\text{stru,ing}}(x) = 0.000401188413745007x^3 - 2.4061095705117x^2 + 4810.17401816086x - 3205416.06613502$	0.70

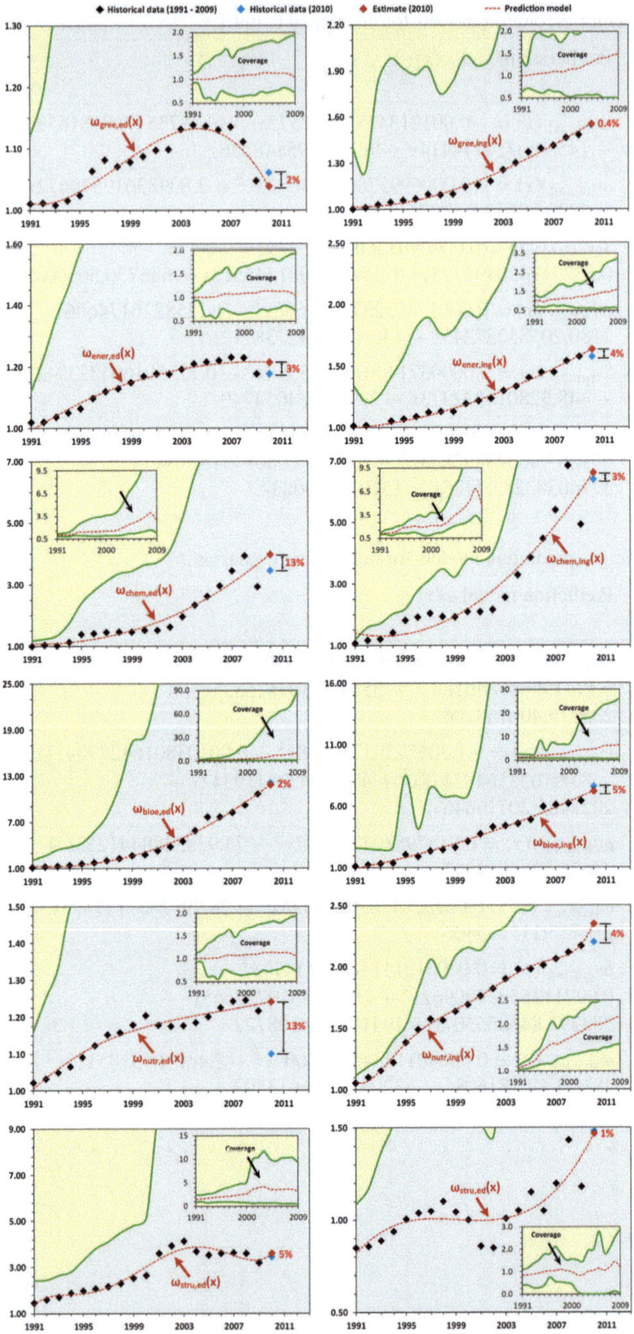

Fig. 5.2 Historical trends and predicted values for each indicator weight for developed and developing countries

In the present study, the method described above was used for model validation. Because the data from 1991 to 2009 were used for the construction of the prediction models, the validation of model accuracy and reliability was performed using historical data from 2010. In the validation results shown in Fig. 5.2, the blue data points represent actual data from 2010, while the red data points represent the values for 2010 predicted by the models. Based on the calculations, the average error of the models is 4.6%, while the minimum and maximum prediction errors are <0.4 and <15%, respectively. Therefore, the models constructed using the time series prediction method have a high accuracy and provide valuable guidance for future work.

5.8 Summary

In this chapter, a method was proposed to determine the weighting system of the assessment system of multi-objective management based on the following aspects definition of weights, selection of the reference year, data mining, selection of countries and regions, construction of absolute weight matrices, testing of the indicator independency, and construction and validation of weight prediction models. Concurrently, the weight data of various indicators were used to perform irreplaceability tests of the indicators. The results indicate that the correlation between energy consumption and GHG emissions indicators is highly significant. However, considering that the two indicators can reflect the environmental impact of the studied subject based on different dimensions and enhance the robustness of the assessment system, both indicators were retained in the indicator system. Subsequently, indicator weight prediction models were constructed for selected developed and developing countries. The average error of the models is 4.6%, which indicates a high prediction accuracy.

Appendix A: Indicator Weights

Table A.1 Weighted coefficients of energy consumption indicator for developed countries during 1990–1999

Country	1990	1991	1992	1993	1994	1995	1996	1997	1998	1999
Canada	1	1.001	1.016	1.023	1.037	1.041	1.058	1.056	1.023	1.065
United States	1	0.986	0.988	0.994	1.000	1.010	1.031	1.025	1.018	1.023
Austria	1	1.037	1.006	1.030	1.018	1.061	1.100	1.112	1.132	1.134
Belgium	1	1.035	1.015	1.012	1.069	1.094	1.157	1.181	1.203	1.181
Denmark	1	1.096	1.026	1.083	1.126	1.128	1.243	1.191	1.160	1.121
Finland	1	1.012	1.055	1.060	1.091	1.037	1.045	1.074	1.113	1.100
France	1	1.061	1.060	1.056	1.050	1.074	1.109	1.098	1.117	1.125
Greece	1	1.005	0.964	1.008	1.016	1.048	1.066	1.114	1.181	1.172
Hungary	1	0.956	0.892	0.873	0.877	0.883	0.916	0.898	0.906	0.907
Iceland	1	0.965	0.977	1.015	1.014	1.027	1.096	1.133	1.231	1.319
Ireland	1	1.039	1.051	1.075	1.122	1.141	1.204	1.266	1.358	1.428
Italy	1	1.022	1.016	1.010	0.994	1.038	1.041	1.054	1.083	1.100
Luxembourg	1	1.070	1.049	1.081	1.027	0.899	0.903	0.893	0.859	0.893
Netherlands	1	1.048	1.032	1.046	1.030	1.045	1.086	1.071	1.062	1.052
Norway	1	0.942	0.991	0.984	0.961	1.016	0.955	0.986	1.009	1.047
Poland	1	0.976	0.964	0.995	0.955	0.925	1.034	1.020	0.960	0.992
Portugal	1	1.001	1.012	1.024	1.070	1.108	1.151	1.193	1.264	1.321
Spain	1	1.054	1.045	1.018	1.056	1.088	1.110	1.180	1.232	1.287
Sweden	1	0.990	0.998	0.992	0.992	1.009	0.991	1.010	1.046	1.030
Switzerland	1	1.023	1.022	1.005	1.013	1.002	0.978	0.993	1.015	1.060
Turkey	1	1.026	1.038	1.116	1.050	1.152	1.250	1.309	1.317	1.257
United Kingdom	1	1.023	0.994	1.024	1.017	1.005	1.066	1.032	1.027	1.028
Israel	1	1.022	1.110	1.056	1.187	1.152	1.189	1.251	1.239	1.223
Australia	1	0.984	0.999	1.015	1.007	1.025	1.058	1.128	1.124	1.165
Japan	1	1.026	1.024	1.031	1.071	1.100	1.116	1.141	1.119	1.142
Korea	1	1.118	1.230	1.358	1.456	1.574	1.656	1.785	1.669	1.794

Table A.2 Weighted coefficients of energy consumption indicator for developed countries during 2000–2009

Country	2000	2001	2002	2003	2004	2005	2006	2007	2008	2009
Canada	1.063	1.038	1.047	1.073	1.089	1.107	1.071	1.081	1.063	0.986
United States	1.035	0.997	1.002	0.997	1.010	1.002	0.986	0.993	0.964	0.910
Austria	1.138	1.182	1.187	1.201	1.226	1.263	1.248	1.231	1.232	1.181
Belgium	1.213	1.196	1.183	1.226	1.236	1.221	1.209	1.205	1.275	1.140
Denmark	1.094	1.106	1.062	1.109	1.063	1.035	1.100	1.069	1.020	0.959
Finland	1.102	1.112	1.128	1.189	1.209	1.133	1.184	1.187	1.155	1.069
France	1.133	1.152	1.136	1.142	1.155	1.152	1.148	1.122	1.124	1.055
Greece	1.236	1.249	1.250	1.319	1.307	1.312	1.349	1.366	1.337	1.284
Hungary	0.885	0.916	0.921	0.944	0.946	1.010	1.003	0.972	0.969	0.906
Iceland	1.382	1.395	1.424	1.428	1.430	1.425	1.545	1.786	2.216	2.181
Ireland	1.499	1.548	1.515	1.476	1.501	1.508	1.554	1.496	1.460	1.316
Italy	1.109	1.113	1.115	1.151	1.166	1.174	1.163	1.145	1.138	1.063
Luxembourg	0.928	0.960	0.997	1.037	1.129	1.138	1.122	1.099	1.066	0.933
Netherlands	1.077	1.108	1.104	1.114	1.138	1.162	1.133	1.137	1.122	1.094
Norway	1.081	1.028	1.058	0.991	1.026	1.063	0.972	1.038	1.037	1.011
Poland	0.905	0.863	0.859	0.893	0.926	0.919	0.964	0.967	0.975	0.999
Portugal	1.374	1.387	1.377	1.414	1.402	1.395	1.356	1.381	1.316	1.303
Spain	1.367	1.414	1.413	1.464	1.491	1.492	1.479	1.490	1.422	1.317
Sweden	0.986	1.043	0.986	0.937	0.995	1.003	0.950	0.963	0.951	0.892
Switzerland	1.032	1.068	1.009	1.017	1.003	0.975	0.988	0.983	1.007	0.999
Turkey	1.347	1.213	1.298	1.348	1.406	1.473	1.583	1.684	1.622	1.507
United Kingdom	1.018	1.029	1.011	1.020	1.021	1.015	0.997	0.954	0.938	0.889
Israel	1.315	1.386	1.308	1.302	1.281	1.287	1.275	1.245	1.159	1.247
Australia	1.158	1.184	1.186	1.182	1.207	1.223	1.215	1.220	1.222	1.200
Japan	1.164	1.151	1.145	1.144	1.175	1.170	1.180	1.167	1.131	1.067
Korea	1.867	1.915	1.975	2.026	2.078	2.145	2.166	2.245	2.281	2.291

Table A.3 Weighted coefficients of energy consumption indicators for developing countries during 1990–1999

Country	1990	1991	1992	1993	1994	1995	1996	1997	1998	1999
Mexico	1	1.047	1.050	1.045	1.068	1.033	1.059	1.073	1.114	1.120
Argentina	1	1.037	1.062	1.009	1.135	1.155	1.183	1.203	1.243	1.243
Barbados	1	1.005	1.011	0.953	0.962	0.982	0.870	1.073	1.102	1.124
Brazil	1	1.026	1.023	0.830	1.080	1.128	1.179	1.224	1.245	1.248
Chile	1	0.981	1.036	1.029	1.154	1.260	1.350	1.506	1.464	1.534
Columbia	1	1.064	1.053	1.058	1.141	1.116	1.195	1.216	1.212	1.140
Saint Lucia	1	0.988	1.008	0.907	1.092	1.095	1.079	1.121	1.111	1.103
Iran	1	1.001	1.008	0.961	1.087	1.115	1.139	1.257	1.281	1.337
Iraq	1	0.679	0.928	0.907	1.148	1.142	1.102	0.990	0.975	0.963
Oman	1	1.100	1.030	1.085	1.115	1.026	1.027	1.157	1.462	1.238
Yemen	1	0.998	1.001	0.718	0.741	0.781	0.771	0.777	0.810	0.822
Algeria	1	1.088	1.019	0.830	0.927	0.957	0.911	0.853	0.875	0.861
Benin	1	1.109	1.205	1.255	1.262	1.339	1.395	1.470	1.525	1.614
Cameroon	1	0.983	0.959	0.708	0.860	0.858	0.834	0.856	0.836	0.798
Egypt	1	0.979	0.962	0.913	1.022	1.029	1.104	1.119	1.131	1.151
Gabon	1	0.998	1.000	0.865	0.920	0.846	0.855	0.878	0.827	0.723
Mali	1	1.006	1.020	1.025	1.024	1.010	0.993	0.996	0.982	0.955
Morocco	1	1.015	1.020	0.929	1.179	1.111	1.166	1.169	1.179	1.224
South Africa	1	1.022	1.025	0.980	1.071	1.063	1.063	1.148	1.080	1.095
Zambia	1	0.937	0.952	0.852	0.804	0.792	0.743	0.791	0.734	0.695
Zimbabwe	1	1.026	0.980	0.872	0.834	0.824	0.837	0.806	0.796	0.897
China	1	1.029	1.057	1.073	1.205	1.215	1.232	1.297	1.263	1.237
Fiji	1	0.847	0.861	0.811	0.927	0.979	1.000	0.987	0.950	1.037
India	1	1.043	1.082	1.049	1.172	1.322	1.253	1.296	1.331	1.397
Malaysia	1	1.080	1.100	1.100	1.313	1.319	1.436	1.424	1.402	1.411
Thailand	1	1.078	1.139	1.208	1.414	1.582	1.813	1.893	1.760	1.795

Table A.4 Weighted coefficients of energy consumption indicators for developing countries during 2000–2009

Country	2000	2001	2002	2003	2004	2005	2006	2007	2008	2009
Mexico	1.156	1.134	1.121	1.133	1.116	1.155	1.238	1.208	1.204	1.148
Argentina	1.260	1.222	1.151	1.226	1.265	1.329	1.388	1.435	1.453	1.423
Barbados	1.238	1.189	1.183	1.199	0.818	1.100	1.044	0.944	0.873	1.008
Brazil	1.271	1.240	1.239	1.238	1.268	1.300	1.330	1.370	1.381	1.361
Chile	1.553	1.584	1.593	1.654	1.758	1.858	1.894	1.538	1.878	1.694
Columbia	1.123	1.103	1.087	1.111	1.108	1.140	1.164	1.183	1.178	1.142
Saint Lucia	1.742	1.815	1.838	2.007	1.930	1.917	1.964	2.058	2.143	2.245
Iran	1.370	1.452	1.570	1.636	1.679	1.900	1.964	1.995	2.021	2.222
Iraq	0.945	0.961	0.942	0.811	0.876	0.899	0.866	0.905	0.920	0.964
Oman	1.370	1.363	1.386	1.344	1.407	1.773	1.997	2.067	2.399	2.533
Yemen	0.862	0.887	0.944	0.990	0.967	0.978	1.037	0.917	1.116	1.200
Algeria	0.839	0.839	0.841	0.840	0.810	0.938	0.965	1.001	1.051	1.109
Benin	1.940	1.896	1.944	2.216	2.501	2.536	2.891	3.046	3.129	3.236
Cameroon	0.811	0.795	0.742	0.778	0.786	0.760	0.801	0.833	0.869	0.783
Egypt	1.175	1.276	1.267	1.337	1.389	1.433	1.322	1.382	1.598	1.640
Gabon	0.644	0.651	0.631	0.633	0.623	0.595	0.642	0.646	0.775	0.649
Mali	0.950	0.930	0.969	0.950	0.954	1.016	0.922	0.918	0.913	0.991
Morocco	1.235	1.312	1.315	1.307	1.294	1.293	1.326	1.341	1.501	1.513
South Africa	1.111	1.113	1.075	1.143	1.211	1.178	1.206	1.247	1.278	1.217
Zambia	0.676	0.694	0.697	0.710	0.707	0.731	0.745	0.722	0.706	0.715
Zimbabwe	0.830	0.777	0.778	0.774	0.693	0.770	0.703	0.693	0.669	0.660
China	1.224	1.286	1.462	1.694	2.073	2.237	2.377	2.508	2.734	2.900
Fiji	1.180	1.264	1.443	1.730	1.876	1.831	1.894	1.869	1.986	2.129
India	1.423	1.449	1.406	1.432	1.523	1.595	1.696	1.784	1.865	1.994
Malaysia	1.558	1.610	1.635	1.664	1.620	1.609	1.633	1.578	1.743	1.767
Thailand	1.828	1.897	2.050	2.228	2.371	2.497	2.514	2.591	2.642	2.639

Table A.5 Weighted coefficients of greenhouse gas emission indicator for developed countries during 1990–1999

Country	1990	1991	1992	1993	1994	1995	1996	1997	1998	1999
Canada	1	0.978	1.003	0.990	0.997	1.012	1.023	1.062	1.062	1.085
United States	1	0.978	0.983	0.989	0.990	0.989	1.012	1.013	1.008	1.007
Austria	1	1.067	0.997	1.013	0.998	1.020	1.115	1.104	1.154	1.110
Belgium	1	1.034	0.986	0.992	1.064	1.088	1.115	1.135	1.174	1.112
Denmark	1	1.120	1.067	1.027	1.119	1.191	1.240	1.272	1.019	0.974
Finland	1	0.958	0.920	0.969	1.068	0.915	0.971	0.983	0.913	0.871
France	1	1.062	1.027	0.986	0.962	0.988	1.031	1.010	1.072	1.051
Greece	1	0.987	0.958	0.997	1.010	1.020	1.029	1.079	1.146	1.128
Hungary	1	0.988	0.938	0.887	0.889	0.870	0.899	0.892	0.885	0.890
Iceland	1	0.932	0.966	1.013	1.033	1.001	1.119	1.124	1.187	1.159
Ireland	1	1.035	1.057	1.072	1.113	1.144	1.190	1.273	1.338	1.392
Italy	1	1.002	0.998	0.987	0.964	1.029	1.008	1.008	1.046	1.045
Luxembourg	1	1.048	1.024	1.069	0.971	0.776	0.776	0.741	0.671	0.703
Netherlands	1	1.028	0.992	1.030	1.023	1.020	1.040	1.082	1.073	1.056
Norway	1	0.926	1.015	0.966	1.015	1.059	1.109	1.156	1.155	1.204
Poland	1	0.983	0.976	1.005	0.954	0.912	1.018	0.994	0.927	0.964
Portugal	1	1.010	1.083	1.039	1.048	1.141	1.083	1.167	1.242	1.394
Spain	1	1.054	1.073	1.021	1.047	1.075	1.047	1.167	1.206	1.303
Sweden	1	0.982	1.019	1.041	1.105	1.089	1.141	1.113	1.120	1.095
Switzerland	1	1.025	1.029	0.970	0.947	0.969	0.985	0.961	0.999	1.003
Turkey	1	1.040	1.027	1.056	1.004	1.080	1.172	1.242	1.234	1.197
United Kingdom	1	1.011	0.953	0.961	0.941	0.918	0.963	0.924	0.913	0.902
Israel	1	0.995	1.111	1.180	1.203	1.174	1.152	1.234	1.244	1.237
Australia	1	0.988	1.000	1.019	1.004	1.019	1.053	1.140	1.148	1.196
Japan	1	1.015	1.019	1.007	1.056	1.051	1.066	1.085	1.040	1.079
Korea	1	1.102	1.185	1.321	1.391	1.497	1.546	1.648	1.461	1.613

Table A.6 Weighted coefficients of greenhouse gas emission indicator for developed countries during 2000–2009

Country	2000	2001	2002	2003	2004	2005	2006	2007	2008	2009
Canada	1.089	1.058	1.059	1.106	1.124	1.137	1.080	1.094	1.067	0.972
United States	1.029	0.999	0.998	0.998	1.009	1.003	0.981	0.988	0.949	0.875
Austria	1.106	1.171	1.219	1.277	1.296	1.332	1.257	1.221	1.200	1.096
Belgium	1.148	1.123	1.098	1.163	1.179	1.158	1.091	1.085	1.184	1.000
Denmark	0.919	0.942	0.890	1.030	0.936	0.860	0.980	0.936	0.890	0.812
Finland	0.908	0.960	0.966	1.200	1.104	0.931	1.037	1.040	0.980	0.933
France	1.041	1.046	1.030	1.041	1.042	1.043	1.042	1.052	1.059	0.972
Greece	1.192	1.211	1.200	1.242	1.228	1.227	1.229	1.271	1.229	1.169
Hungary	0.848	0.880	0.883	0.918	0.895	0.919	0.905	0.877	0.869	0.780
Iceland	1.219	1.151	1.187	1.182	1.218	1.206	1.227	1.378	1.306	1.202
Ireland	1.451	1.524	1.469	1.411	1.451	1.446	1.457	1.393	1.374	1.206
Italy	1.059	1.050	1.065	1.112	1.103	1.109	1.098	1.079	1.056	0.960
Luxembourg	0.735	0.769	0.825	0.859	0.950	0.962	0.938	0.913	0.879	0.769
Netherlands	1.096	1.156	1.117	1.125	1.148	1.159	1.159	1.103	1.062	1.054
Norway	1.120	1.121	1.133	1.216	1.283	1.117	1.086	1.110	1.051	1.138
Poland	0.864	0.817	0.811	0.843	0.865	0.852	0.887	0.877	0.874	0.903
Portugal	1.389	1.329	1.429	1.334	1.361	1.413	1.310	1.267	1.180	1.189
Spain	1.373	1.378	1.434	1.448	1.510	1.538	1.486	1.506	1.377	1.239
Sweden	1.024	1.015	0.961	1.021	1.012	0.964	0.948	0.943	0.912	0.844
Switzerland	0.983	0.987	0.947	0.956	0.959	0.960	0.951	0.913	0.940	0.947
Turkey	1.310	1.176	1.223	1.276	1.284	1.388	1.487	1.637	1.573	1.462
United Kingdom	0.904	0.920	0.899	0.913	0.925	0.919	0.918	0.887	0.872	0.798
Israel	1.263	1.375	1.318	1.275	1.280	1.310	1.279	1.266	1.204	1.216
Australia	1.185	1.215	1.215	1.206	1.229	1.244	1.240	1.247	1.245	1.227
Japan	1.119	1.109	1.113	1.158	1.163	1.148	1.147	1.162	1.127	1.026
Korea	1.659	1.689	1.743	1.771	1.798	1.821	1.781	1.846	1.909	1.938

Table A.7 Weighted coefficients of greenhouse gas emission indicator for developing countries during 1990–1999

Country	1990	1991	1992	1993	1994	1995	1996	1997	1998	1999
Mexico	1	1.007	0.999	0.992	1.035	0.972	0.990	1.024	1.072	1.036
Argentina	1	1.045	1.041	1.121	1.076	1.093	1.173	1.162	1.199	1.221
Barbados	1	1.003	1.028	1.028	0.961	1.006	0.788	1.078	1.096	1.118
Brazil	1	1.019	0.969	1.037	1.064	1.126	1.176	1.229	1.206	1.231
Chile	1	0.942	0.958	0.998	1.081	1.165	1.300	1.518	1.510	1.636
Columbia	1	1.132	1.243	1.252	1.200	1.183	1.227	1.336	1.350	1.198
Saint Lucia	1	0.988	1.022	0.991	1.096	1.098	1.086	1.127	1.116	1.108
Iran	1	1.076	1.083	1.086	1.136	1.174	1.161	1.265	1.265	1.348
Iraq	1	0.637	0.846	0.946	1.063	1.033	1.001	0.881	0.884	0.898
Oman	1	1.096	1.005	1.056	1.046	0.952	0.956	1.145	1.356	1.246
Yemen	1	0.991	0.993	0.789	0.746	0.779	0.773	0.741	0.758	0.793
Algeria	1	1.008	0.942	0.923	0.924	0.946	0.893	0.839	0.867	0.851
Benin	1	1.111	1.223	1.279	1.287	1.365	1.425	1.502	1.571	1.656
Cameroon	1	0.983	0.989	1.819	1.980	1.919	1.685	1.674	1.631	1.602
Egypt	1	1.000	0.982	0.985	1.004	0.998	1.076	1.088	1.091	1.093
Gabon	1	0.979	1.008	0.963	0.913	0.842	0.831	0.820	0.786	0.741
Mali	1	1.011	1.013	1.021	1.015	0.996	1.001	0.990	0.975	0.948
Morocco	1	1.017	1.012	1.110	1.195	1.134	1.149	1.117	1.116	1.185
South Africa	1	1.020	1.031	1.006	1.076	1.062	1.063	1.143	1.073	1.092
Zambia	1	0.870	1.031	0.820	0.640	0.615	0.684	0.720	0.655	0.562
Zimbabwe	1	1.038	1.034	0.962	0.927	0.894	0.884	0.821	0.811	0.882
China	1	1.030	1.052	1.116	1.190	1.190	1.192	1.259	1.203	1.162
Fiji	1	0.809	0.837	0.885	0.924	0.978	1.007	0.993	0.949	1.043
India	1	1.053	1.098	1.128	1.177	1.369	1.279	1.320	1.351	1.409
Malaysia	1	1.005	1.063	1.099	1.197	1.211	1.342	1.314	1.287	1.299
Thailand	1	1.092	1.169	1.312	1.446	1.619	1.876	1.926	1.753	1.829

Table A.8 Weighted coefficients of greenhouse gas emission indicator for developing countries during 2000–2009

Country	2000	2001	2002	2003	2004	2005	2006	2007	2008	2009
Mexico	1.077	1.054	1.055	1.056	1.012	1.053	1.143	1.149	1.157	1.093
Argentina	1.193	1.096	1.026	1.126	1.170	1.247	1.277	1.357	1.363	1.315
Barbados	1.256	1.209	1.203	1.220	0.837	1.100	1.066	0.982	1.056	1.052
Brazil	1.244	1.241	1.217	1.197	1.215	1.246	1.275	1.315	1.384	1.329
Chile	1.485	1.453	1.405	1.493	1.652	1.727	1.783	1.527	1.632	1.552
Columbia	1.186	1.150	1.114	1.133	1.092	1.159	1.162	1.168	1.206	1.225
Saint Lucia	1.781	1.853	1.876	2.043	1.937	1.961	1.961	2.116	2.217	2.207
Iran	1.347	1.384	1.495	1.575	1.642	1.794	1.874	1.899	1.962	2.077
Iraq	0.851	0.873	0.869	0.760	0.810	0.958	0.900	0.942	0.971	0.982
Oman	1.291	1.286	1.303	1.248	1.326	1.639	1.865	1.937	2.247	2.422
Yemen	0.822	0.843	0.901	0.944	0.916	0.927	0.982	0.932	1.115	1.272
Algeria	0.830	0.774	0.787	0.777	0.756	0.836	0.867	0.915	0.955	0.991
Benin	1.997	1.933	1.981	2.257	2.554	2.601	2.999	3.188	3.153	3.193
Cameroon	1.543	1.427	1.331	1.286	1.329	1.361	1.443	1.502	1.432	1.374
Egypt	1.091	1.151	1.163	1.222	1.276	1.314	1.221	1.250	1.404	1.424
Gabon	0.673	0.640	0.594	0.603	0.593	0.581	0.515	0.508	0.524	0.497
Mali	0.949	0.924	0.985	0.968	0.977	1.055	0.933	0.935	0.951	0.986
Morocco	1.177	1.257	1.239	1.251	1.191	1.227	1.250	1.251	1.276	1.270
South Africa	1.106	1.113	1.065	1.144	1.214	1.176	1.198	1.238	1.288	1.208
Zambia	0.529	0.557	0.563	0.576	0.564	0.601	0.554	0.525	0.511	0.507
Zimbabwe	0.779	0.743	0.708	0.638	0.563	0.622	0.586	0.559	0.533	0.520
China	1.141	1.182	1.372	1.603	1.995	2.149	2.257	2.387	2.582	2.754
Fiji	1.256	1.325	1.517	1.952	2.265	2.150	1.917	2.081	1.941	1.804
India	1.444	1.452	1.415	1.414	1.518	1.571	1.677	1.763	1.872	2.032
Malaysia	1.396	1.439	1.542	1.599	1.569	1.564	1.564	1.534	1.657	1.772
Thailand	1.711	1.805	1.946	2.122	2.308	2.456	2.392	2.475	2.534	2.555

Table A.9 Weighted coefficients of greenhouse gas emission indicator for developed countries during 1990–1999

Country	1990	1991	1992	1993	1994	1995	1996	1997	1998	1999
Canada	1	1.039	1.097	1.213	1.377	1.547	1.640	1.821	1.852	1.936
United States	1	1.029	1.148	1.217	1.392	1.605	1.677	1.818	1.855	2.038
Austria	1	0.998	0.997	0.997	0.995	0.995	0.941	0.894	0.903	0.858
Belgium	1	0.988	1.035	1.116	1.234	1.580	1.557	1.592	1.654	1.636
Denmark	1	0.961	1.039	0.872	0.984	1.234	1.131	1.110	1.068	1.019
Finland	1	0.809	0.810	0.773	0.962	1.162	1.096	1.129	1.104	1.041
France	1	0.942	1.007	0.883	0.995	1.230	1.194	1.148	1.190	1.187
Greece	1	1.041	1.163	1.067	1.093	1.638	1.393	1.321	1.417	1.280
Hungary	1	1.009	1.023	1.002	1.280	1.483	1.500	1.646	1.976	2.059
Iceland	1	1.064	0.991	0.891	0.922	1.090	1.159	1.115	1.212	1.218
Ireland	1	1.039	1.082	1.049	1.219	1.512	1.641	1.810	1.648	1.649
Italy	1	0.992	1.027	0.916	1.029	1.269	1.262	1.252	1.251	1.236
Luxembourg	1	1.032	1.050	1.070	1.077	1.090	1.103	1.132	1.135	1.259
Netherlands	1	0.888	0.947	0.939	0.986	1.224	1.174	1.120	1.075	1.012
Norway	1	0.995	1.022	0.951	1.095	1.350	1.321	1.277	1.255	1.191
Poland	1	1.176	1.231	1.340	1.743	2.388	2.792	3.106	3.350	3.435
Portugal	1	1.006	1.129	0.904	1.036	1.311	1.295	1.278	1.303	1.269
Spain	1	1.020	1.091	0.898	1.092	1.456	1.447	1.478	1.563	1.544
Sweden	1	0.937	0.967	1.003	1.103	1.283	1.233	1.213	1.297	1.309
Switzerland	1	0.934	0.963	0.892	0.996	1.196	1.172	1.154	1.160	1.071
Turkey	1	0.953	1.023	1.132	0.987	1.626	1.701	1.839	1.759	1.591
United Kingdom	1	0.969	1.013	0.928	1.053	1.354	1.292	1.302	1.293	1.259
Israel	1	0.987	0.971	0.956	0.923	0.913	0.884	0.854	0.841	0.842
Australia	1	0.987	1.059	1.100	1.269	1.420	1.581	1.486	1.421	1.471
Japan	1	1.074	1.004	1.049	1.204	1.462	1.417	1.442	1.276	1.379
Korea	1	1.035	1.066	1.106	1.133	1.519	1.500	1.486	1.026	1.234

Table A.10 Weighted coefficients of greenhouse gas emission indicator for developed countries during 2000–2009

Country	2000	2001	2002	2003	2004	2005	2006	2007	2008	2009
Canada	2.059	2.023	2.057	2.242	2.563	2.862	3.108	3.209	3.353	2.677
United States	2.436	2.455	2.436	2.701	3.076	3.528	3.713	3.748	4.284	3.125
Austria	0.886	0.880	0.923	1.223	1.336	1.532	1.704	1.977	2.232	1.680
Belgium	1.714	1.745	1.871	2.159	2.679	3.206	3.862	4.365	4.502	3.509
Denmark	0.984	0.960	1.089	1.233	1.406	1.540	1.726	1.887	2.117	1.567
Finland	1.068	1.032	1.094	1.332	1.572	1.800	2.018	2.296	2.529	1.706
France	1.161	1.122	1.150	1.393	1.593	1.748	1.853	2.191	2.373	1.761
Greece	1.243	1.212	1.611	1.842	2.198	2.282	2.501	2.970	3.385	2.555
Hungary	2.060	2.054	2.246	2.737	3.350	3.627	4.148	5.045	6.002	4.388
Iceland	1.185	1.417	1.531	1.653	1.916	2.009	2.193	2.663	2.806	1.916
Ireland	1.718	1.632	1.702	2.012	2.338	2.376	2.478	2.616	2.621	1.984
Italy	1.278	1.207	1.266	1.519	1.816	1.916	2.110	2.438	2.500	1.928
Luxembourg	1.166	1.348	1.275	1.636	1.881	1.993	2.217	2.042	2.307	1.650
Netherlands	1.081	1.002	1.039	1.341	1.617	1.854	2.075	2.428	2.812	2.039
Norway	1.156	1.172	1.196	1.412	1.629	1.893	2.166	2.502	2.907	2.419
Poland	3.638	3.721	4.110	5.124	6.528	7.553	8.858	11.065	13.998	10.363
Portugal	1.260	1.251	1.343	1.613	1.855	2.056	2.237	2.600	2.838	2.344
Spain	1.516	1.517	1.676	2.067	2.384	2.526	2.743	3.124	3.340	2.540
Sweden	1.236	1.201	1.193	1.509	1.761	1.897	2.119	2.518	2.956	2.228
Switzerland	1.058	1.062	1.192	1.417	1.528	1.637	1.767	2.113	2.200	1.808
Turkey	1.854	1.461	1.841	2.350	3.139	3.667	4.109	4.975	5.512	4.146
United Kingdom	1.327	1.269	1.304	1.540	1.827	1.933	2.086	2.686	2.495	1.970
Israel	0.960	0.894	0.883	0.954	1.185	1.167	1.320	1.477	1.656	1.207
Australia	1.336	1.255	1.288	1.728	2.013	2.294	2.377	2.733	3.404	2.533
Japan	1.609	1.521	1.504	1.756	2.095	2.346	2.569	2.812	3.387	2.632
Korea	1.454	1.346	1.428	1.650	2.076	2.464	2.739	3.131	3.456	2.763

Table A.11 Weighted coefficients of chemical use indicator for developing countries during 1990–1999

Country	1990	1991	1992	1993	1994	1995	1996	1997	1998	1999
Mexico	1	1.195	1.592	1.716	2.028	1.995	2.317	2.760	2.947	3.226
Argentina	1	1.060	1.120	1.171	1.445	1.721	2.083	2.277	2.217	1.993
Brazil	1	1.070	1.045	1.282	1.553	2.188	2.363	2.603	2.556	2.315
Chile	1	1.194	1.294	1.305	1.401	1.831	1.990	2.000	1.859	1.735
Latvia	1	1.033	1.067	1.133	1.179	2.133	2.781	3.147	3.482	3.207
Russia	1	1.029	1.059	1.059	1.118	1.118	1.123	1.238	1.026	0.781
South Africa	1	1.125	1.232	1.226	1.354	1.668	1.559	1.530	1.308	1.318
China	1	1.000	0.994	0.850	1.066	1.509	1.595	1.681	1.688	1.908
India	1	0.971	1.045	0.927	1.313	1.673	1.494	1.682	1.568	1.695
Malaysia	1	1.096	1.236	1.312	1.522	2.065	1.866	1.878	1.363	1.585
Thailand	1	1.011	1.182	1.302	1.576	2.049	1.918	1.629	1.270	1.395

Table A.12 Weighted coefficients of chemical use indicator for developing countries during 2000–2009

Country	2000	2001	2002	2003	2004	2005	2006	2007	2008	2009
Mexico	3.783	3.735	3.900	4.222	4.870	5.552	6.159	6.604	7.285	5.712
Argentina	2.035	1.787	1.207	1.653	2.203	2.499	2.745	3.513	4.415	2.792
Brazil	2.459	2.389	2.177	2.426	3.230	3.326	3.673	5.018	7.542	5.158
Chile	1.900	1.921	1.744	2.010	2.495	2.978	3.365	3.860	5.291	3.558
Latvia	3.276	3.560	3.985	5.161	6.912	8.033	10.312	13.059	16.279	9.638
Russia	0.881	1.133	1.340	1.702	2.157	2.870	3.759	5.128	6.443	5.073
South Africa	1.369	1.237	1.368	1.593	1.959	2.206	2.405	2.806	3.399	2.441
China	2.399	2.495	3.044	3.812	5.055	5.906	6.525	7.891	8.491	7.802
India	1.425	1.450	1.578	1.814	2.318	3.306	3.819	4.753	7.882	5.975
Malaysia	1.868	1.639	1.651	1.689	2.362	2.489	2.734	3.110	3.518	2.712
Thailand	1.668	1.598	1.729	2.016	2.562	3.136	3.509	3.612	4.651	3.195

Table A.13 Weighted coefficients of bioenergy recovery indicator for developed countries during 1990–1999

Country	1990	1991	1992	1993	1994	1995	1996	1997	1998	1999
Canada	1	0.994	1.099	1.170	1.374	1.335	1.367	1.594	1.680	1.886
United States	1	1.064	1.155	1.174	1.182	1.166	1.168	1.173	1.162	1.165
Austria	1	1.051	1.135	1.149	1.016	1.551	1.464	1.496	1.552	1.425
Belgium	1	1.112	1.193	1.195	1.203	1.415	1.421	1.215	1.355	1.618
Denmark	1	1.691	2.476	3.444	3.781	4.271	5.466	6.301	6.786	8.287
Finland	1	0.964	0.950	1.138	1.229	1.252	1.160	1.541	1.821	1.630
France	1	1.104	1.126	1.048	1.156	1.310	1.427	1.609	1.593	1.855
Hungary	1	1.408	2.674	2.274	2.666	3.056	2.706	2.833	3.349	3.599
Italy	1	2.116	1.783	2.009	2.727	3.742	5.810	7.878	11.793	17.434
Luxembourg	1	1.106	0.918	1.132	1.144	1.457	1.138	1.231	1.188	1.380
Netherlands	1	1.107	1.070	1.235	1.345	1.462	1.962	2.231	2.412	2.653
Norway	1	0.975	1.014	1.114	1.205	1.259	1.324	1.058	1.171	1.186
Poland	1	1.692	1.633	1.402	1.368	1.393	1.545	2.305	2.281	1.934
Portugal	1	1.173	1.279	1.302	1.341	1.414	1.369	1.470	1.439	1.734
Spain	1	0.997	1.053	1.043	1.205	1.950	2.229	3.064	3.167	3.582
Sweden	1	0.951	1.020	1.094	1.109	1.171	1.057	1.367	1.466	1.388
Switzerland	1	1.019	1.122	1.105	1.254	1.295	1.390	1.635	1.770	1.915
United Kingdom	1	1.141	1.522	1.996	2.730	2.989	3.223	3.753	4.671	5.730
Australia	1	1.012	0.870	0.860	0.852	0.906	1.156	1.191	1.251	1.357
Japan	1	1.028	1.048	1.029	1.087	1.179	1.241	1.368	1.388	1.434

Table A.14 Weighted coefficients of bioenergy recovery indicator for developed countries during 2000–2009

Country	2000	2001	2002	2003	2004	2005	2006	2007	2008	2009
Canada	1.852	1.922	2.003	1.989	1.922	1.985	1.869	1.797	1.505	1.618
United States	1.173	0.951	1.018	1.002	0.996	1.001	1.002	1.004	0.986	0.968
Austria	1.336	1.520	1.397	1.555	1.869	2.302	3.016	3.669	3.799	3.874
Belgium	1.795	2.121	2.222	2.166	2.594	2.993	4.125	4.831	5.822	6.967
Denmark	8.513	9.613	11.396	14.350	16.104	17.976	17.480	17.456	17.827	18.141
Finland	1.628	1.565	1.695	1.774	1.971	1.791	2.021	1.866	1.947	1.647
France	2.080	2.507	2.646	2.800	2.872	2.847	2.792	3.102	3.220	3.331
Hungary	3.607	3.794	2.142	5.864	22.745	52.472	42.390	51.956	62.469	74.746
Italy	18.210	24.627	33.398	42.674	53.475	58.324	63.903	65.876	72.605	88.407
Luxembourg	1.438	1.494	1.524	1.554	1.875	1.827	2.161	2.416	2.549	2.726
Netherlands	2.845	3.135	3.624	3.402	4.014	5.756	5.651	4.702	5.600	6.444
Norway	1.116	1.289	1.221	1.643	1.720	1.458	1.707	1.716	1.751	1.061
Poland	2.110	2.929	3.408	2.836	4.523	6.702	8.546	10.690	14.679	20.974
Portugal	2.164	2.218	2.393	2.294	2.477	2.708	2.731	2.909	2.891	3.208
Spain	3.030	2.802	4.035	4.862	4.931	4.159	4.448	4.708	5.149	5.374
Sweden	2.087	1.929	2.127	2.313	3.818	3.982	4.451	5.062	5.323	5.781
Switzerland	2.043	2.129	2.174	2.180	2.267	2.394	2.637	2.591	2.680	2.654
United Kingdom	6.378	10.047	11.288	12.674	13.418	16.318	16.234	15.394	15.211	16.918
Australia	1.341	1.582	1.829	1.853	2.103	4.280	4.314	4.304	4.951	4.221
Japan	1.410	1.415	1.527	1.649	1.675	2.035	2.055	2.119	2.070	1.982

Table A.15 Weighted coefficients of bioenergy recovery indicator for developing countries during 1990–1999

Country	1990	1991	1992	1993	1994	1995	1996	1997	1998	1999
Mexico	1	1.015	1.408	1.533	1.501	1.772	1.066	0.927	1.062	0.961
Argentina	1	0.986	0.973	0.960	1.138	1.124	3.240	4.118	4.252	4.831
Brazil	1	1.020	1.230	1.224	1.311	1.340	1.593	1.714	1.717	1.886
Chile	1	1.055	1.751	1.730	1.739	1.803	1.645	1.627	1.070	0.919
Russia	1	0.995	0.993	0.941	0.862	0.846	0.835	0.823	0.818	1.120
Gabon	1	2.605	2.536	2.469	2.704	2.632	1.994	1.942	1.892	1.831
China	1	1.000	1.000	1.000	1.000	1.041	1.149	2.760	4.307	3.627
Fiji	1	1.000	1.000	1.000	1.018	8.506	4.121	7.517	6.768	6.836
India	1	0.991	0.983	0.974	0.966	0.957	0.950	0.941	0.933	0.926
Malaysia	1	1.045	1.045	1.136	1.182	2.182	2.591	2.591	2.591	4.462
Thailand	1	1.000	1.000	2.105	2.110	2.115	3.155	3.190	4.060	4.725

Table A.16 Weighted coefficients of bioenergy recovery indicator for developing countries during 2000–2009

Country	2000	2001	2002	2003	2004	2005	2006	2007	2008	2009
Mexico	0.984	1.458	1.422	1.392	1.410	1.703	1.277	1.279	1.321	1.438
Argentina	6.017	5.434	7.564	8.617	10.233	10.961	23.350	22.767	22.686	22.447
Brazil	1.743	1.967	2.207	2.533	2.621	2.817	3.012	3.641	3.913	4.437
Chile	0.846	1.839	1.709	1.572	1.749	1.526	1.208	2.254	2.554	3.510
Russia	1.376	1.563	1.535	0.997	1.002	1.470	1.534	1.124	1.438	1.501
Gabon	1.770	1.725	1.683	1.643	1.604	1.568	1.535	1.503	1.473	1.445
China	5.584	5.521	4.607	4.559	4.566	4.575	4.584	4.593	4.604	4.629
Fiji	6.842	6.852	6.793	6.735	6.678	6.621	6.561	6.505	6.383	6.344
India	0.919	0.912	0.906	0.898	0.891	0.884	0.876	0.868	0.859	0.852
Malaysia	6.084	8.084	8.040	8.010	8.010	8.012	7.998	7.990	7.982	7.976
Thailand	6.250	10.615	17.050	17.800	18.390	21.065	23.000	24.065	24.925	26.368

Table A.17 Weighted coefficients of nutrient recovery indicator for developed countries during 1990–1999

Country	1990	1991	1992	1993	1994	1995	1996	1997	1998	1999
United States	1	0.985	0.976	0.979	0.976	0.978	0.959	0.990	0.998	1.014
Austria	1	0.964	0.967	1.008	1.021	1.041	1.132	1.170	1.171	1.224
Belgium	1	1.016	1.065	1.097	1.143	1.307	1.302	1.340	1.321	1.341
Denmark	1	1.111	1.250	1.333	1.496	1.447	1.603	1.509	1.515	1.603
Finland	1	1.013	1.018	1.028	1.032	1.033	1.025	1.120	1.166	1.212
France	1	1.018	1.040	1.064	1.067	1.073	1.095	1.118	1.142	1.143
Greece	1	1.006	1.046	1.043	1.040	1.033	1.160	1.254	1.310	1.366
Hungary	1	0.990	0.982	0.979	0.918	0.872	0.889	0.925	0.920	0.916
Iceland	1	1.005	1.041	1.057	1.080	1.095	1.118	1.138	1.154	1.168
Ireland	1	1.032	1.056	1.076	1.080	1.105	1.127	1.173	1.192	1.240
Italy	1	1.160	1.223	1.298	1.286	1.277	1.284	1.313	1.323	1.397
Luxembourg	1	0.959	0.853	0.865	0.832	1.003	0.999	1.030	1.067	1.100
Netherlands	1	0.998	1.032	1.116	1.127	1.102	1.131	1.184	1.190	1.202
Norway	1	1.037	1.100	1.091	1.157	1.325	1.337	1.310	1.368	1.259
Poland	1	0.955	0.951	0.950	0.981	0.977	1.033	1.083	1.051	1.094
Portugal	1	1.045	1.089	1.182	1.253	1.276	1.311	1.329	1.383	1.404
Spain	1	1.058	1.089	1.182	1.253	1.276	1.311	1.329	1.383	1.404
Sweden	1	1.030	1.080	1.115	1.145	1.178	1.237	1.296	1.306	1.418
Switzerland	1	1.016	1.024	1.046	1.067	1.076	1.075	1.111	1.164	1.145
Turkey	1	0.995	0.976	0.978	0.975	0.978	0.986	0.995	1.010	1.050
United Kingdom	1	1.029	1.051	1.061	1.120	1.169	1.239	1.325	1.345	1.223
Israel	1	1.012	1.019	1.032	1.042	1.052	1.080	1.124	1.144	1.201
Australia	1	1.008	1.023	1.025	1.042	1.044	1.121	1.189	1.126	1.160
Japan	1	1.014	1.006	1.011	1.019	1.024	1.035	1.036	1.044	1.044
Korea	1	1.087	0.875	0.725	0.664	0.541	0.561	0.534	0.493	0.501

Table A.18 Weighted coefficients of nutrient recovery indicator for developed countries during 2000–2009

Country	2000	2001	2002	2003	2004	2005	2006	2007	2008	2009
United States	1.030	1.008	1.017	1.014	1.034	1.023	1.026	1.013	0.988	0.949
Austria	1.266	1.259	1.326	1.329	1.355	1.367	1.427	1.427	1.437	1.422
Belgium	1.378	1.364	1.414	1.360	1.420	1.406	1.424	1.467	1.462	1.470
Denmark	1.695	1.685	1.712	1.659	1.721	1.834	1.845	1.961	2.063	1.933
Finland	1.258	1.164	1.148	1.166	1.176	1.199	1.243	1.277	1.319	1.220
France	1.160	1.189	1.201	1.147	1.178	1.204	1.219	1.234	1.229	1.215
Greece	1.422	1.455	1.478	1.497	1.516	1.536	1.557	1.577	1.599	1.621
Hungary	0.846	0.857	0.866	0.878	0.860	0.871	0.884	0.863	0.857	0.813
Iceland	1.189	1.200	1.223	1.237	1.285	1.325	1.468	1.481	1.478	1.484
Ireland	1.288	1.508	1.494	1.575	1.584	1.564	1.696	1.660	1.541	1.393
Italy	1.423	1.442	1.463	1.469	1.522	1.546	1.586	1.588	1.584	1.567
Luxembourg	1.113	1.098	1.107	1.149	1.151	1.143	1.165	1.189	1.201	1.178
Netherlands	1.236	1.235	1.250	1.226	1.253	1.248	1.240	1.256	1.240	1.217
Norway	1.301	0.765	0.832	0.855	0.882	0.909	0.984	1.060	1.061	1.033
Poland	1.086	0.987	0.934	0.883	0.869	1.084	1.090	1.094	1.088	1.076
Portugal	1.450	1.433	1.457	1.481	1.466	1.485	1.528	1.544	1.695	1.698
Spain	1.450	1.433	1.457	1.481	1.466	1.485	1.528	1.544	1.695	1.698
Sweden	1.523	1.513	1.490	1.521	1.497	1.469	1.475	1.446	1.379	1.356
Switzerland	1.143	1.181	1.252	1.254	1.247	1.298	1.341	1.404	1.406	1.331
Turkey	1.085	1.081	1.109	1.088	1.098	1.099	1.158	1.182	1.219	1.197
United Kingdom	1.209	1.206	1.187	1.172	1.104	1.147	1.085	1.080	0.998	0.970
Israel	1.216	1.247	1.263	1.249	1.272	1.230	1.235	1.203	1.148	1.111
Australia	1.300	1.051	1.152	1.189	1.280	1.248	1.147	1.296	1.135	1.012
Japan	1.064	1.059	1.047	1.047	1.029	1.016	1.003	0.980	0.929	0.905
Korea	0.506	0.525	0.537	0.544	0.534	0.515	0.518	0.533	0.550	0.536

Table A.19 Weighted coefficients of nutrient recovery indicator for developing countries during 1990–1999

Country	1990	1991	1992	1993	1994	1995	1996	1997	1998	1999
Mexico	1	1.015	1.039	1.304	1.344	1.369	1.411	1.272	1.308	1.308
Brazil	1	0.986	0.974	0.957	0.946	0.940	0.937	0.934	0.934	0.931
Chile	1	1.026	1.046	1.062	1.106	1.144	1.158	1.166	1.258	1.286
Russia	1	1.024	1.068	1.152	1.165	1.524	1.817	2.059	1.905	2.087
China	1	1.114	1.191	1.253	1.403	1.489	1.496	1.505	1.537	1.542

Table A.20 Weighted coefficients of nutrient recovery indicator for developing countries during 2000–2009

Country	2000	2001	2002	2003	2004	2005	2006	2007	2008	2009
Mexico	1.281	1.296	1.308	1.322	1.374	1.389	1.401	1.413	1.425	1.436
Brazil	0.933	0.889	0.890	0.894	0.907	0.909	0.820	0.758	0.742	0.733
Chile	1.340	1.356	1.385	1.413	1.439	1.440	1.492	1.506	1.555	1.573
Russia	2.154	2.135	2.269	2.327	2.457	2.464	2.612	2.711	2.800	2.912
China	1.587	1.799	1.813	1.963	2.039	2.037	1.931	1.970	1.989	2.017

Table A.21 Weighted coefficients of struvite recovery indicator for developed countries during 1990–1999

Country	1990	1991	1992	1993	1994	1995	1996	1997	1998	1999
United States	1	1.024	0.986	0.736	0.840	0.880	0.908	0.907	0.863	0.784
Finland	1	0.860	1.033	1.131	1.179	1.217	1.162	1.225	1.270	1.299
Israel	1	2.426	2.447	2.650	2.672	2.456	2.360	2.380	2.518	3.202
Australia	1	1.500	1.900	2.300	2.800	3.300	3.800	4.100	4.500	4.800

Table A.22 Weighted coefficients of struvite recovery indicator for developed countries during 2000–2009

Country	2000	2001	2002	2003	2004	2005	2006	2007	2008	2009
United States	0.737	0.603	0.676	0.649	0.658	0.661	0.543	0.530	0.534	0.463
Finland	1.327	1.352	1.406	1.402	1.467	1.439	1.497	1.448	1.358	1.148
Israel	3.344	2.571	3.196	2.841	2.473	2.388	2.137	2.184	2.160	1.855
Australia	5.129	9.819	10.376	11.566	10.080	9.567	10.168	10.270	10.267	9.232

Table A.23 Weighted coefficients of struvite recovery indicator for developing countries during 1990–1999

Country	1990	1991	1992	1993	1994	1995	1996	1997	1998	1999
Mexico	1	0.939	0.698	0.761	0.815	0.913	0.984	1.014	1.058	1.280
Brazil	1	1.032	0.880	1.049	1.176	1.144	1.107	1.212	1.245	1.200
Chile	1	0.933	1.099	0.898	0.606	0.728	1.025	0.734	1.806	1.159
Algeria	1	0.944	0.967	0.604	0.597	0.600	0.817	0.813	0.870	0.812
Egypt	1	0.418	0.393	0.288	0.285	0.398	0.837	1.082	1.067	0.988
Morocco	1	0.825	0.864	0.812	0.887	0.871	0.878	0.960	0.928	0.921
South Africa	1	0.962	0.934	0.732	0.743	0.803	0.754	0.766	0.757	0.807
Zimbabwe	1	0.770	0.912	0.961	0.933	0.951	0.745	0.562	0.541	0.497
China	1	1.084	1.171	0.947	1.190	1.367	1.347	1.080	1.157	1.235
India	1	0.820	0.697	1.359	1.525	1.782	1.790	2.114	1.633	1.513

Table A.24 Weighted coefficients of struvite recovery indicator for developing countries during 2000–2009

Country	2000	2001	2002	2003	2004	2005	2006	2007	2008	2009
Mexico	1.434	1.059	0.006	0.007	0.000	0.000	0.010	0.060	1.200	1.741
Brazil	1.287	1.258	1.345	1.457	1.465	1.431	1.488	1.531	1.646	1.474
Chile	1.039	1.053	1.072	1.156	1.153	1.083	0.728	1.308	2.129	0.681
Algeria	0.641	0.676	0.526	0.634	0.557	0.616	1.019	1.200	1.189	0.662
Egypt	1.041	0.725	1.109	1.945	1.935	2.790	1.819	2.048	2.547	2.911
Morocco	0.784	0.857	0.896	0.897	1.013	1.080	1.005	1.020	0.901	0.656
South Africa	0.754	0.646	0.740	0.690	0.707	0.660	0.667	0.642	0.570	0.554
Zimbabwe	0.452	0.502	0.492	0.466	0.488	0.270	0.392	0.277	0.127	0.194
China	0.948	1.022	1.084	1.141	1.197	1.250	1.592	1.847	2.053	2.424
India	1.562	1.521	1.450	1.689	1.995	2.336	1.782	2.046	1.967	1.663

References

BP (2006) Statistic review of World Energy British Petroleum Company, London.

Brockwell, P. J., Davis, R. A., & Calder, M. V. (2002) Introduction to time series and forecasting, Springer.

IEA. (2004). World Energy Outlook, International Energy Agency, Paris.

UNEP. (2012). UNEP Year Book 2012, United Nations Environment Programme.

Chapter 6
Scenario Analysis for the Multi-objective Management of Municipal Wastewater

6.1 Overview

The ultimate purpose of model construction and simulation is not merely prediction; more importantly, they enable the simulation and assessment of policies formulated by policymakers. Within this context, a scenario analysis method has been proposed. A scenario is not a single prophecy or prediction; it provides plausible directions and trends for future development. During scenario design, each designer conceives and envisages the future blueprint, or more precisely, develops a future development concept. Therefore, a retrospective analysis of the objective development history is required before scenario definition and construction. Subsequently, a set of reasonable, generally accepted, and justifiable assumptions or confirmed future prospects should be developed. Finally, the possibilities of achieving the prospects and measures that must be adopted are analyzed (Khan et al. 2017). The greatest difference between scenario analysis and conventional prediction methods is that the scenario analysis involves the investigation of future situations that may occur under certain presuppositions and the conditions that lead to these situations, while prediction merely attempts to describe situations that are most likely to occur in the future (Wang et al. 2018).

In Chap. 4, we proposed the following concept of multi-objective management of municipal wastewater treatment. Given that the effluent quality meets certain discharge standards, the energy and material consumptions required for wastewater treatment and environmental footprint (e.g., greenhouse gas (GHG) emissions) are reduced to the greatest extent possible and existing recovery technologies are utilized for the recovery of organic matter and nutrients, such as nitrogen and phosphorus, from wastewater and excess sludge via multiple routes to achieve the management of multiple objectives such as energy conservation, reduced carbon emissions, and resource recovery (see Sect. 4.2 for details). This concept is the core of the present chapter and the backbone of the entire study. To satisfy the needs of the scenario

X. Wang, *Energy Consumption, Chemical Use and Carbon Footprints of Wastewater Treatment Alternatives*, Springer Theses,
https://doi.org/10.1007/978-981-13-5983-5_6

analysis in this chapter, environmental footprint, bioenergy recovery, nutrient recovery, and simultaneous bioenergy and nutrient recovery scenarios were defined based on the technological levels of existing wastewater and sludge treatment and disposal methods during the construction of Chap. 4 (Construction of an assessment system for multi-objective management of municipal wastewater treatment). Therefore, based on the preceding chapters, the indicator weights of various scenarios were used as starting points to conduct scenario analysis and predictions for multi-objective management of wastewater treatment to determine reasonable management methods for future municipal wastewater treatment and provide a strong basis for decision-making by relevant authorities. The use of the weight prediction models, which were constructed in Chap. 5, for the scenario analysis and predictions for multi-objective management of wastewater treatment, is the main novelty of this chapter.

6.2 Scenario Analysis

6.2.1 Scenario Analysis Concept

The concepts of the scenarios and scenario analysis were briefly mentioned in Chap. 4 but not discussed in detail. The term "scenario" first appeared in the book "The Year 2000" written by Herman Kahn and Anthony Wiener in 1967. The authors asserted that the future possesses diversity and that several potential outcomes may be achieved. More than one path leads to any of the possible outcomes and the descriptions of the plausible future and associated paths constitute a scenario. Therefore, a scenario is the description of a future situation and the succession of events that leads to the development of a situation from the initial state to the future state. This includes the qualitative and quantitative descriptions of the basic characteristics of various states and descriptions of the potential occurrence of various states (Balarezo and Nielsen 2017).

Scenario analysis refers to the process in which various plausible future schemes are conceived through detailed and rigorous deductions and descriptions of the future based on key assumptions with respect to major economic, industrial, or technological changes. The greatest advantage of scenario analysis is that it enables the management personnel and decision-makers to discover future development trends related to certain situations, thereby avoiding the over- or underestimation of future changes and impacts (Postma and Liebl 2005; Rockafellar and Wets 1991; Swart et al. 2004; Yang and Lin 2016). Accordingly, scenario analysis has the following characteristics: (1) future development trends are diverse, that is, many trends possibly occur and develop, and predicted future outcomes will therefore also be diverse; (2) it includes the acknowledgment of subjective initiatives of decision-makers with respect to future development, considering population prospects of decision-makers as important aspects of scenario analysis and unimpeded information exchanges with decision-makers during scenario analysis; (3) the key factors that emphasize the coordination consistency and play a key role in organizational development are analyzed during the scenario analysis; (4) the value of scenario analysis lies in the fact that it enables decision-makers to prepare for a certain event and development trends of the event and to take action to minimize negative factors and maximize positive factors.

6.2.2 *Steps for Scenario Analysis*

The scenario analysis steps include scenario preparation, scenario field analysis, scenario prognostics, scenario development, and scenario transfer (Amer et al. 2013; Balarezo and Nielsen 2017; Postma and Liebl 2005; Wiek et al. 2006). In this chapter, the following steps have been devised by combining the scenario analysis steps formulated by the Stanford Research Institute (SRI) with the research needs of the present study:

(1) Determination of the core components of the decision-making process: the core components of the decision-making process must include two elements: importance and uncertainty. Decision-makers should focus on a limited number of critical issues such as the respective objectives involved in multi-objective municipal wastewater treatment, that is, the reductions of the energy consumption and GHG emissions and resource recovery. The core issues of the scenario analysis, which is an important analytical technique that is used for the prediction of future trends, must possess a certain degree of uncertainty, that is, different outcomes may be generated.

(2) Construction of scenarios: by combining the assumptions of different management objectives, different scenarios are constructed to provide a basis for the decision-making by relevant authorities. In this chapter, the concept of multi-objective management of wastewater treatment was used as the basis for the construction of several scenarios, that is, the environmental footprint, bioenergy recovery, nutrient recovery, and simultaneous bioenergy and nutrient recovery scenarios.

(3) Selection of analysis indicators: analysis indicators represent the concretization of various key factors. In general, analysis indicators are obtained through brainstorming of plausible future situations and prediction of prospects by experts or subject populations during seminars or other events. The participants of such events are usually grouped based on the primary influencing factors and are invited to provide opinions about certain issues. This leads to the generation of a series of viewpoint inventories (i.e., a series of assessment indicators), which are subsequently categorized and sorted. However, this indicator-selection method involves a certain degree of subjectivity. During the selection of the assessment indicators in the present work, a preliminary selection was performed first based on literature research. Subsequently, the indicator irreplaceability was tested to determine the final assessment indicators (see Sects. 4.5 and 5.6).

(4) Sorting of the importance and uncertainty of analysis indicators: during conventional scenario analysis, the levels of the importance of the driving factors (i.e., assessment indicators) are subjectively defined based on the levels of the impact of the factors on the decision-making process. In this chapter, relevant analysis was performed based on subjective and objective approaches: firstly, it was assumed that the analysis indicators have equivalent importance, that is, the impacts of various driving factors on the decision-making process are identical. Secondly, subjective weights were assigned to the assessment indicators by utilizing the weight prediction models constructed in Chap. 5.
(5) Analysis of the scenario content: based on different assumptions, the continuity of the management of wastewater treatment in different scenarios was analyzed. In addition, future development trends were predicted, which serve as a theoretical basis for the management personnel and decision-makers.

6.3 Scenario Analysis Methods for the Multi-objective Management of Wastewater Treatment

6.3.1 Scenario Design Assumptions

In Sect. 4.3, we proposed four scenarios: environmental footprint scenario, bioenergy recovery scenario, nutrient recovery scenario, and simultaneous bioenergy and nutrient recovery scenarios. The premises and assumptions used during the design of these scenarios, which were not stated previously, are discussed in detail in the following section:

(1) Resource recovery is one of the objectives of the management of wastewater treatment proposed in the present study. The recovery of organic matter and nutrients, such as nitrogen and phosphorus, in wastewater and excess sludge is primarily considered. If excess sludge contains high concentrations of toxic substances, such as heavy metals, the resource recovery of excess sludge may lead to potential ecotoxicity problems. However, the sustainability has not been discussed from the ecotoxicology perspective in the present work. Therefore, the scenario analysis focuses on municipal wastewater treatment processes and systems of the study subject. It is assumed that toxic substances, such as heavy metals, in wastewater and sludge do not have an ecological impact on the receiving environments.
(2) In this chapter, scenario analysis of the multiple objectives of sustained management of municipal wastewater treatment was performed to forecast future trends of the sustained management of wastewater and predict the best management scenarios. Therefore, typical combinations of anaerobic–anoxic–aerobic treatment processes were selected in the present study, and it was assumed that

various treatment systems have identical water qualities and geological environments. The various treatment systems used in this study have been discussed in Sect. 3.3.

(3) The literature review showed that the current development of secondary biological processes for wastewater treatment faces great challenges, mainly based on high energy consumption, excess sludge production, and GHG emissions during the wastewater treatment and the fact that there is an energy shortage and the GHG emissions need to be reduced. The energy consumption, chemical consumption, and GHG emissions during wastewater treatment are significantly associated with the fossil fuel consumption and environmental problems such as global warming, ecotoxicity, and acidification. Therefore, a macro-approach was adopted for the environmental footprint scenario. The energy consumption, chemical consumption, and GHG emissions inventory of the wastewater treatment process were used as the quantification indicators of the environmental footprint, rather than using the refinement indicators of the respective environmental impacts for scenario analysis. In addition, during the scenario analysis, excess sludge was not regarded to be waste with an environmental footprint. Instead, the excess sludge was considered to be a recyclable resource. It is the main subject of the design for the bioenergy recovery, nutrient recovery, and simultaneous bioenergy and nutrient recovery scenarios.

(4) The main routes of resource recovery from wastewater and excess sludge in the resource recovery scenarios constructed in the present study are the methanogenesis during the anaerobic fermentation of sludge, sludge composting, and recovery of struvite from supernatant produced during sludge fermentation. Regarding other popular wastewater and sludge resource recovery routes, such as biological hydrogen production through fermentation and microbial fuel cells, several issues remain with respect to the actual industrialization processes. Therefore, these popular resource recovery routes were not included in the scenario design to ensure that a clear theoretical basis is provided to the decision-makers.

6.3.2 Approaches to Scenario Construction

The scenario construction process mainly consists of two parts. In the first part, the basic scenario is designed. Based on the assumptions described above, the wastewater treatment process is described, which includes the selection of the wastewater treatment system, construction of the process model, and numerical simulation of the process. Based on the proposed concept of multi-objective management of wastewater treatment, the environmental footprint of the wastewater and sludge treatment processes are quantified in the second part and multiple resource recovery routes and processes are specified from the viewpoint of the resource recovery of sludge. Figure 6.1 shows the schematic diagram of the scenario for the multi-objective management of wastewater treatment.

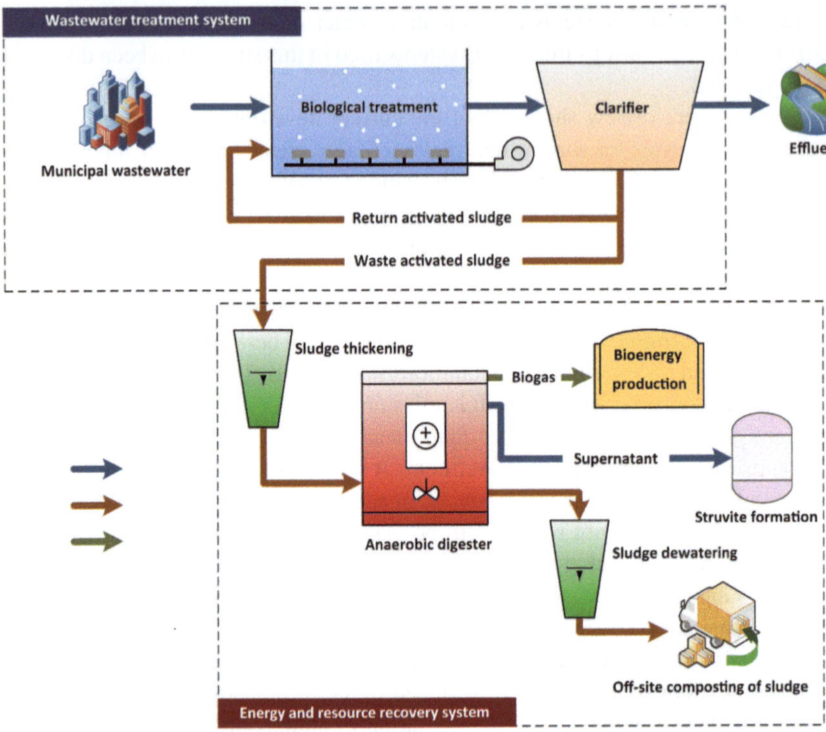

Fig. 6.1 Schematic diagram of the scenario for the multi-objective management of wastewater treatment

(1) Environmental footprint scenario

Based on the assumption that the environmental issues associated with environments in which wastewater treatment plants (WWTPs) are located become increasingly sensitive in the future due to the impact of various factors, the quantification and assessment of the environmental footprint contributed by WWTPs and the formulation of effective counter-strategies are necessary. Therefore, environmental burdens are primarily considered in the environmental footprint scenario. The research approach in Chap. 3 (Environmental impact inventory analysis of typical wastewater treatment processes) was used as reference for the definition and construction of this scenario. Energy consumption, chemical consumption, and GHG emissions were used to represent the environmental footprint inventory.

(2) Bioenergy recovery scenario

Based on the assumption of the severe depletion of fossil fuels and increasing prominence of environmental impacts of traditional fossil fuels (e.g., coal and petroleum), the self-supply or self-compensation of the energy sources in WWTPs will become inevitable. The utilization of organic matter in wastewater or sludge and the transformation of such matter into energy sources form the core of the bioenergy recovery scenario. Therefore, biogas (primarily CH_4) generated during the anaerobic digestion of excess sludge is collected for energy recovery through biogas combustion in this scenario. The environmental footprint of the treatment and disposal of wastewater and sludge, including the energy recovery process, is also considered in this scenario, in addition to the potential energy recovery.

(3) Nutrient recovery scenario

In this scenario, it is assumed that future WWTPs produce a huge amount of excess sludge and the pollution transfer problems arising from sludge disposal methods, such as sludge incineration, have become increasingly severe. To increase the recovery and utilization rates of the resources in waste material and to replenish the soil fertility, the excess sludge produced by the WWTP is thickened and dewatered before transportation out of the plant for nutrient recovery via natural composting methods. In addition, the following points must be noted: (1) the excess sludge mainly originates from the municipal WWTP; (2) the ecotoxic impacts of residual heavy metals in the excess sludge are neglected; (3) in addition to investigating the nitrogen and phosphorus contents of the excess sludge, only the environmental footprint (energy consumption, chemical consumption, and GHG emissions) of the wastewater and sludge treatment and disposal was taken into account.

(4) Simultaneous bioenergy and nutrient recovery scenario

Based on the assumption that the background problems of scenarios (1) and (3) simultaneously occur and become increasingly severe, the main purpose of this scenario is to enhance the bioenergy and nutrient recovery and management while considering methods to reduce the environmental footprint. In the simultaneous bioenergy and nutrient recovery scenario, CH_4 produced from excess sludge of the WWTP during anaerobic digestion will be collected for the energy recovery via combustion; simultaneously, nitrogen and phosphorus in the supernatant produced from anaerobic digestion of sludge are recovered in the form of struvite, while the anaerobically digested sludge is thickened, dewatered, and subsequently transported out of the plant for agricultural use via natural composting. The environmental footprint of both the wastewater treatment and resource recovery, including the resource recovery potential of pollutants, is considered in this scenario.

6.3.3 Technical Scenario Analysis Roadmap

Different research methods are used for different subjects during scenario analysis, that is, from the quantitative descriptions of the scenarios and core scenario parameters to the scenario predictions. For instance, computer platforms were used for simulations and data mining to obtain basic data on the construction and operation of wastewater treatment systems. The indicator system of the environmental footprint scenario was preliminarily determined based on the literature review. Subsequently, the credibility of the indicator system was enhanced by testing the indicator irreplaceability based on historical weight data. During the later stage of the scenario analysis, the weight prediction models developed in Chap. 5 were utilized for subjective weight-based scenario predictions of the multi-objective management of wastewater treatment. The detailed technical scenario analysis roadmap is shown in Fig. 6.2.

6.4 Simulation of the Weight-Based Multi-objective Management of Wastewater Treatment

6.4.1 Analytical Methods

The scenario analysis in this chapter was performed based on the assessment system for the multi-objective management of wastewater treatment constructed in Chap. 4. Prior to the scenario analysis, the following four steps were performed: (1) delineation of assessment boundaries; (2) selection of the indicator system; (3) acquisition of assessment data; and (4) characterization of assessment indicators. The process flowchart is shown in Fig. 6.2, and the details of the respective steps are provided in Sect. 4.5.

6.4.2 Study Objective

The study objective of this chapter differs from that of Chap. 3 in that the main foci of Chap. 3 are the secondary biological treatment systems of municipal WWTPs. Six common combinations of anaerobic–anoxic–aerobic treatment processes were selected as study subjects for the analysis and comparison of the environmental impact of different processes. By contrast, wastewater and sludge treatment and disposal were used as the basic settings for the investigation of the impacts of more stringent wastewater discharge standards on the environmental footprint, bioenergy recovery, nutrient recovery, and simultaneous bioenergy and nutrient recovery scenarios. The six wastewater treatment systems constructed in Chap. 3 were also used in this chapter. However, a macro-approach was adopted in the scenario analysis to determine the average wastewater treatment levels; horizontal comparisons of the various wastewater treatment systems were not performed.

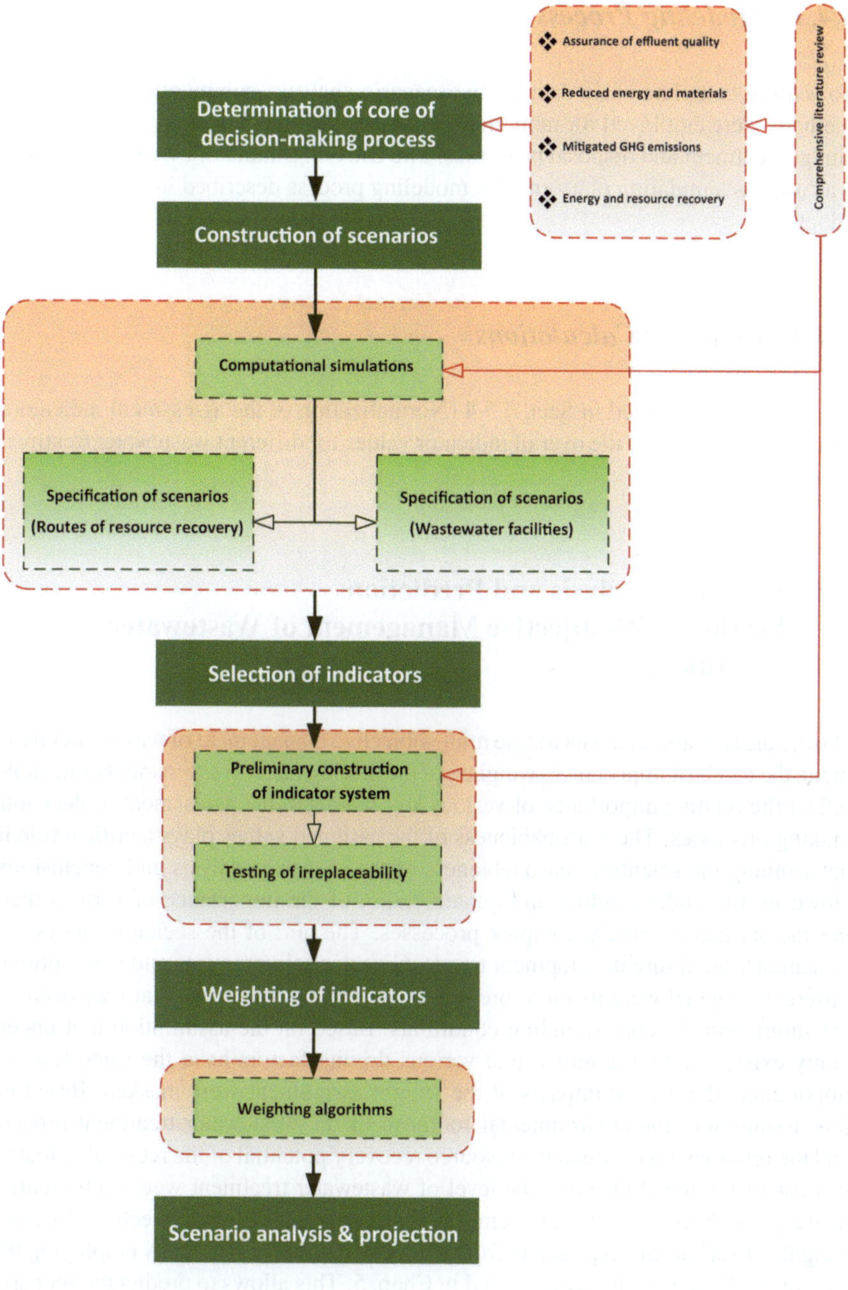

Fig. 6.2 Technical scenario analysis roadmap

6.4.3 Modeling Process

To acquire complete basic data for the scenario analysis, computational simulation methods were employed for numerical simulations of the municipal wastewater and sludge treatment and disposal processes. The BioWin Simulator (V.3.0, BW3-1952) was used as simulation platform; the modeling process described in Sect. 3.3.3 was adopted.

6.4.4 Indicator Calculations

The equations provided in Sect. 4.5.4 (Normalization of the assessment indicators) were used to calculate the overall indicator values for different wastewater treatment scenarios.

6.5 Scenario Analysis and Predictions for the Multi-objective Management of Wastewater Treatment

During the scenario analysis for the multi-objective management of wastewater treatment, the levels of importance (weights) of the driving factors (assessment indicators) reflect the relative importance of various key factors in the assessment or decision-making processes. The reasonableness of the assigned values plays a critical role in determining the scientific reasonableness of the scenario analysis and conclusions. However, the understanding and quantification of the importance of various driving factors are extremely complex processes. The aim of the scenario analysis is to demonstrate future development trends. Given certain premises and assumptions, different assigned weights therefore reflect the various situations that may occur in the future and the corresponding conditions. Based on the assumption that uncertainty exists, we first assumed that various driving factors have the same level of importance, that is, the impacts of the weight assignment were masked. Based on this assumption, the environmental footprint of the wastewater treatment process and the relationship between the resource recovery potential of the recyclable matter in wastewater and sludge and the level of wastewater treatment were preliminarily investigated. Subsequently, an attempt was made to elucidate the effects of different weights of various driving factors from a developmental viewpoint by employing the weight prediction models constructed in Chap. 5. This allows to predict the scenario effects of the enhancement of the wastewater treatment level, which can trigger the contemplation of the management modes for future wastewater treatment.

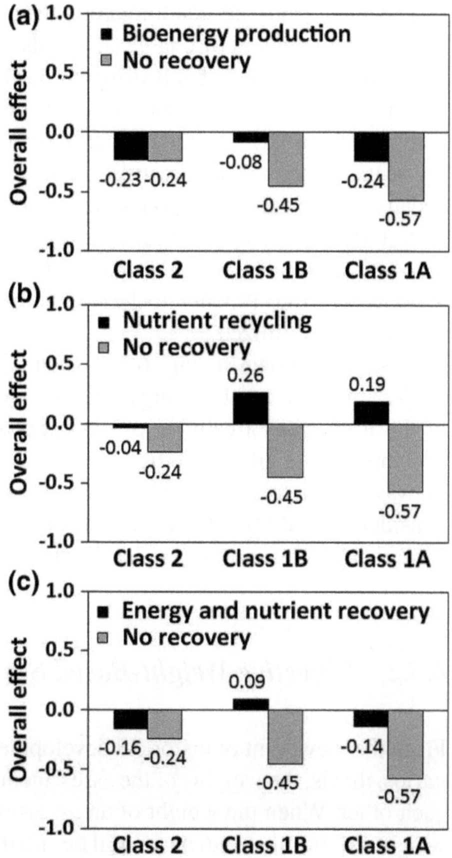

Fig. 6.3 Overall benefits of the wastewater treatment process: **a** bioenergy recovery scenario, **b** nutrient recovery scenario, and **c** simultaneous bioenergy and nutrient recovery scenarios. [Adapted from Wang et al. 2012, with permission from American Chemical Society, Copyright (2012)]

6.5.1 Subjective Weight-Based Scenario Analysis

In the beginning of the study, it was assumed that the assessment indicators have the same level of importance, that is, all indicators have identical weights. Therefore, a weight value of 1 was assigned to each indicator for the preliminary scenario analysis. The results are shown in Fig. 6.3.

Figure 6.3a shows that the overall benefits of the wastewater treatment process of the bioenergy recovery scenario increase by 4.2–82.2% compared with the environmental footprint scenario. However, the overall benefits of this scenario are still influenced by environmental burdens because substantial amounts of energy and chemicals are input into the anaerobic sludge digestion unit. In addition, an interesting phenomenon was observed: when the wastewater treatment complies with a more stringent discharge standard (e.g., Class 1A) in the bioenergy recovery scenario, the overall benefit is identical to that of using a more lenient discharge standard. In other words, from the standpoint of eutrophication control and energy recovery, the adoption of stringent discharge standards may be considered.

Figure 6.3b shows that a stringent discharge standard favors the nutrient recovery in the respective scenario, which involves the recovery of nutrients through the composting of excess sludge. Furthermore, compared with the bioenergy recovery scenario, the nutrient recovery scenario leads to higher overall benefits. This is because the former scenario requires resource consumption in the sludge digestion unit. There are no advantages in using a lenient discharge standard in the nutrient recovery scenario because nutrients, such as nitrogen and phosphorus, are discharged together with the effluent into the receiving waters and therefore cannot be effectively recovered and utilized.

Lastly, as shown in Fig. 6.3c, when a moderate discharge standard is adopted in the simultaneous bioenergy and nutrient recovery scenario, the overall benefit of the wastewater treatment process is positive, indicating a higher sustainability. This provides a greater reference value for the future multi-objective management of wastewater treatment. Therefore, the resource recovery of pollutants must be considered in addition to the impacts of the wastewater on the receiving waters when formulating relevant wastewater discharge standards.

6.5.2 *Objective Weight-Based Scenario Predictions*

From the viewpoint of historical development, the driving factors of the various scenarios, that is, the weights of the assessment factors mutually influence and constrain each other. When the weight of an assessment indicator is set to a larger value, the weights of the other indicators will be smaller. Changes of the weights may also alter the assessment outcomes and lead to changes of the predictions or decisions. Therefore, objective weight-based scenario predictions were performed in this section.

6.5.2.1 Objective Weights of Assessment Indicators

The absolute weights of the assessment indicators for developed and developing countries were calculated in Chap. 5 and absolute weight prediction models were designed. However, subsequent scenario predictions required the relative weights of various indicators, which were calculated using the following equation:

$$\omega'_j(x) = \frac{\omega_j(x)}{\sum_j \omega_j(x)} \tag{6.1}$$

where $\omega'_j(x)$ is the relative weight of the assessment indicator j in the target year (x) and $\omega_j\ (x)$ is the absolute weight of the assessment indicator j in the target year (x).

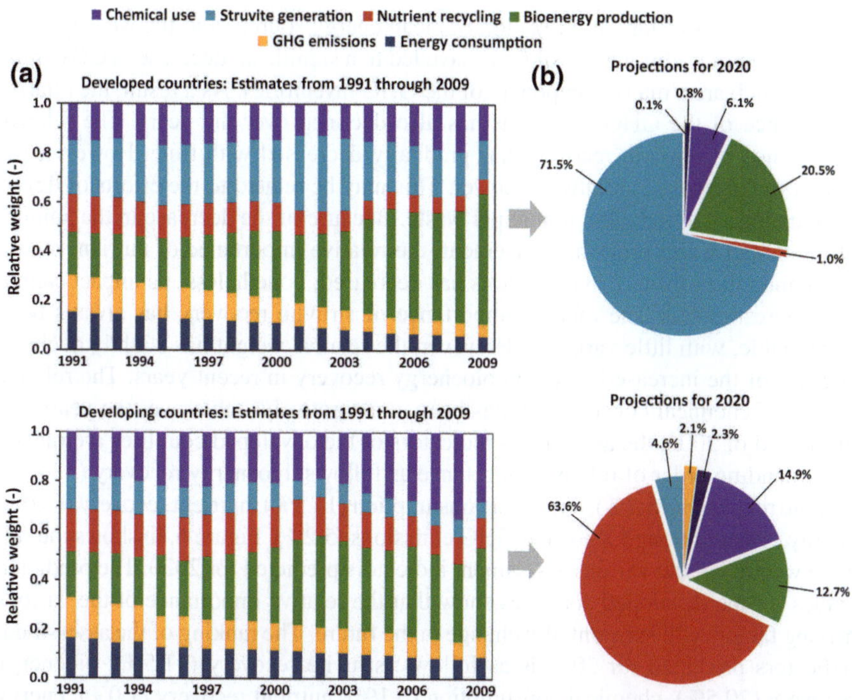

Fig. 6.4 Relative weights of various indicators for developed and developing countries: **a** historical data from 1991 to 2009, **b** predicted data for 2020. (Adapted from Wang et al. 2015)

The relative weights of the various indicators in different years for developed and developing countries were calculated using Eq. (6.1), as shown in Fig. 6.4a. The absolute weights of the indicators in 2020 were obtained using the weight prediction models and the predicted relative weights were subsequently determined using Eq. (6.1), as shown in Fig. 6.4b.

Figure 6.4 shows that there are huge differences in the relative weights of each indicator in different countries, regions, and historical periods. In addition, as discussed previously, the relative weights of various indicators are mutually constraining. If the weight of a certain indicator increases, the weights of the other indicators decrease. Generally speaking, the control of GHG emissions has significantly improved because countries around the world place greater emphasis on the reduction of GHG emissions, which is manifested in a decline of the weight of GHG emissions over the years.

For developed countries, the gradual decrease in the relative weight of the energy consumption and gradual increase in the relative weight of the bioenergy recovery may be related to the energy policies of these countries, which place emphasis on the development and utilization of new and clean energy sources, particularly power generation from organic matter. The trend of the relative weight of GHG emissions shows that the substantial utilization of clean energy sources and the weakening of the position of traditional fossil fuels have led to a significant decrease in CO_2 emissions, which are a major component of the GHG inventories. As a result, the relative importance of the GHG emissions has also declined over the years. The relative importance of nutrient recovery has gradually decreased with time. However, the rate of the decrease is relatively slower. This may be related to the efforts of developed countries in reducing municipal waste. Because of the decrease in the amount of municipal waste requiring placement, the relative importance of nutrient recovery among the various driving factors has decreased; nonetheless, its impact cannot be underestimated. The relative importance of struvite recovery has always been appreciable, with little variation. However, the relative weight has slightly declined because of the increased focus on bioenergy recovery in recent years. The relative weight of chemical consumption has been comparatively stable over the years. As of the end of 2009, the assessment indicators of the developed countries are ranked in descending order of relative importance as follows: bioenergy recovery (52.8%), struvite recovery (15.7%), chemical consumption (14.5%), nutrient recovery (6.0%), energy consumption (5.8%), and GHG emissions (5.2%). Figure 6.4b shows the relative weights of the various assessment indicators predicted for 2020. The predicted values for the developed countries show that the relative importance of the various driving factors will substantially change in the future. The ranking of the assessment indicators predicted for 2020 is as follows: struvite recovery (71.5%), bioenergy recovery (20.5%), chemical consumption (6.1%), nutrient recovery (1.0%), energy recovery (0.8%), and GHG emissions (0.1%). Therefore, struvite recovery will be the issue of greatest concern in developed countries in the future.

In developing countries, the relative weight of the energy consumption has decreased over time. The relative importance of GHG emissions has also gradually declined over the years, which is similar to the trend for developed countries. However, compared with the developed countries, the energy consumption has a higher relative importance in developing countries. In addition, in contrast to the developed countries, developing countries exhibit a significant upward trend in the relative weight of the chemical consumption. This may be related to the differences in the degree of industrialization between developing and developed countries because developing countries still face a huge demand for chemicals in many fields. Similarly, differences also exist between developing and developed countries with respect to the struvite recovery. The relative weight of the struvite recovery in developing countries has decreased over the years, which may be related to the fact that most of the countries with the most commercial phosphate rock reserves are developing countries. The relative importance of the nutrient recovery has always played an important and stable role, which indicates that the generation and placement of municipal waste remain issues of major concern in developing countries. As of the end of 2009, the assessment indicators of developing countries are ranked in descending order of relative importance as follows: bioenergy recovery (36.1%), chemical consumption (27.8%), nutrient recovery (12.2%), energy consumption (8.9%), GHG emissions (8.7%), and struvite recovery (8.4%). The ranking of the assessment indicators for developing countries predicted for 2020 is as follows: nutrient recov-

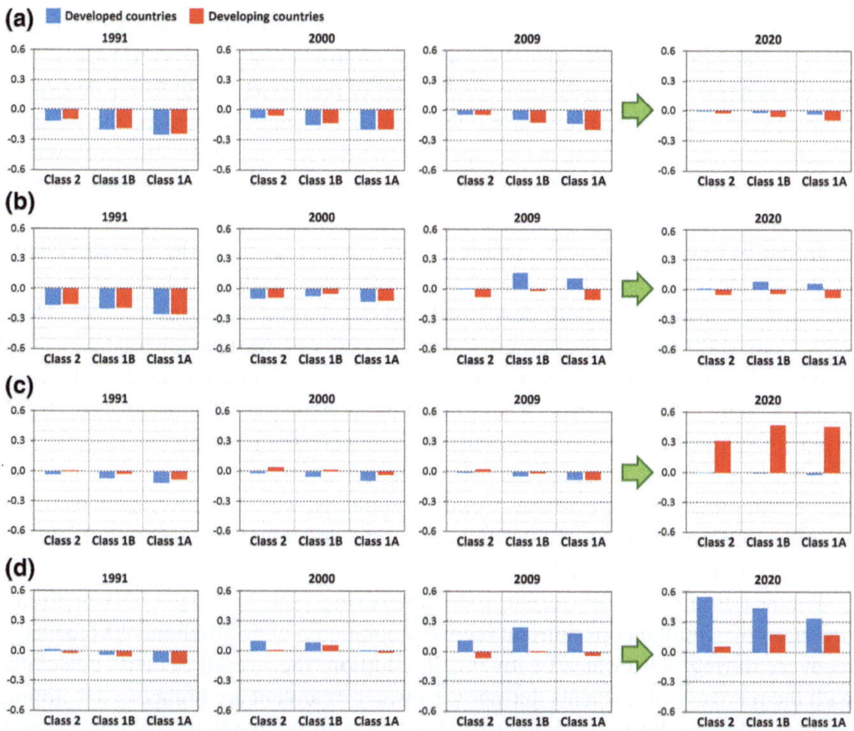

Fig. 6.5 Analysis of the overall benefits of the wastewater treatment process: **a** environmental footprint scenario, **b** bioenergy recovery scenario, **c** nutrient recovery scenario, and **d** simultaneous bioenergy and nutrient recovery scenario

ery (63.6%), chemical consumption (14.9%), bioenergy recovery (12.7%), struvite recovery (4.6%), energy consumption (2.3%), and GHG emissions (2.1%). This shows that the nutrient recovery via sludge composting will be the issue of greatest concern for developing countries in the future.

6.5.2.2 Scenario Analysis and Predictions for the Sustained Management of Wastewater Treatment

In this section, the overall benefits of municipal wastewater treatment of different scenarios for developed and developing countries were analyzed and future situations were predicted. The results are shown in Fig. 6.5.

In the environmental footprint scenario, as shown in Fig. 6.5a, the overall benefits of wastewater treatment are dominated by negative effects and decrease as the wastewater discharge standard becomes more stringent in both developed and developing countries. However, from the standpoint of time series development, the magnitude of the negative effects decreases over time, that is, from 1991 to 2009, which includes almost two decades, although the overall benefits of wastewater treatment remained negative. In particular, a significant improvement is observed between the overall benefits in 1991 and the overall benefits predicted for 2020 for developed countries. For developing countries, the rate of the improvement of the negative effects of wastewater treatment over the past two decades has been slower. However, based on model predictions, the rate of the improvement will increase over the next 10 years.

In the bioenergy recovery scenario, as shown in Fig. 6.5b, the overall benefits of wastewater treatment decrease as the wastewater discharge standard becomes increasingly stringent in both developed and developing countries in 1991. However, from 2000 onwards, optimum overall benefits of wastewater treatment were achieved based on a moderate wastewater discharge standard instead of the most stringent standard. This result is similar to the conclusion obtained in Sect. 6.5.1. From 2009 onwards, the overall benefits of wastewater treatment for developed countries became positive; however, the data predicted for 2020 indicate a declining trend of the magnitude of the positive overall benefits. However, the overall benefits of simultaneous energy recovery during wastewater treatment are generally higher in developed countries than in developing countries.

For the nutrient recovery scenario, as shown in Fig. 6.5c, higher overall benefits are produced by simultaneous nutrient recovery compared with simultaneous bioenergy recovery during wastewater treatment. In addition, the overall benefits associated with the recovery of nutrients during wastewater treatment are higher in developing countries than in developed countries. According to the predicted data, the overall benefits of such a measure in developed countries will be highly positive in 2020, with the highest benefit obtained by the adoption of a moderate wastewater discharge standard.

Figure 6.5d shows that the overall benefits of simultaneous bioenergy and nutrient recovery during wastewater treatment in 1991 decreased as the wastewater discharge standard became increasingly stringent. In particular, the overall benefits for developing countries are slightly higher than that of developed countries. By 2000, the overall benefits were dominated by positive effects and the situation in developed countries significantly changed. Consequently, the overall benefits increased as the wastewater discharge standard became increasingly stringent. However, higher overall benefits were still obtained with a moderate discharge standard in developing countries. In 2009, the overall benefits were dominated by positive effects in developed countries, while the overall benefits of developing countries were more negatively affected. The model-predicted data show that the overall benefits for both developed and developing countries are expected to be highly positive in 2020. However, the overall benefits for developing countries exhibit an increasing trend with an increasingly stringent wastewater discharge standard, while the opposite is true for developed countries, that is, the overall benefits decrease as the wastewater discharge standard becomes more stringent.

6.6 Summary

In this chapter, the weights of the assessment indicators were used as starting points to perform subjective and objective weight-based scenario analysis and predict the multi-objective management of wastewater treatment. The study results are as follows:

(1) When the concept of the resource recovery of pollutants was introduced into the wastewater treatment assessment framework, the negative effects of the wastewater treatment process did not increase with increasingly stringent wastewater discharge standard. The positive environmental effects of the resource recovery of pollutants compensates for the environmental effects of the wastewater treatment.
(2) Huge differences exist in the relative weights of each indicator in different countries, regions, and historical periods. The relative weights of the various indicators are mutually constraining. If the weight of a certain indicator increases, the weights of the other indicators decrease.
(3) Because countries around the world place greater emphasis on the reduction of GHG emissions, the control of GHG emissions has gradually improved. The relative weight of bioenergy recovery was approximately 40% in the past two decades, which is substantial. However, adjustments in the energy structures have become more significant in the recent decade, which is manifested in an increase in the relative weights of the bioenergy recovery. Furthermore, the trend of the relative weight of the bioenergy recovery over the past decade in developing countries is similar to that from 1990 to 2000 in developed countries, which shows that the trends of certain indicators for developing countries follow that of developed countries.
(4) The model-predicted data show that the overall benefits of simultaneous bioenergy and nutrient recovery during wastewater treatment in both developed and developing countries are expected to be highly positive in 2020. The issues of greatest concern in developed and developing countries are the phosphorus recovery via struvite and nutrient recovery via sludge composting, respectively.

References

Amer, M., Daim, T. U., & Jetter, A. (2013). A review of scenario planning. *Futures, 46,* 23–40.
Balarezo, J., & Nielsen, B. B. (2017). Scenario planning as organizational intervention an integrative framework and future research directions. *Review of International Business and Strategy, 27*(1), 2–52.
Khan, Z., Linares, P., & Garcia-Gonzalez, J. (2017). Integrating water and energy models for policy driven applications. A review of contemporary work and recommendations for future developments. *Renewable and Sustainable Energy Reviews, 67,* 1123–1138.
Postma, T., & Liebl, F. (2005). How to improve scenario analysis as a strategic management tool? *Technological Forecasting and Social Change, 72*(2), 161–173.

Rockafellar, R. T., & Wets, R. J. B. (1991). Scenarios and policy aggregation in optimization under uncertainty. *Mathematics of Operations Research, 16*(1), 119–147.

Swart, R. J., Raskin, P., & Robinson, J. (2004). The problem of the future: sustainability science and scenario analysis. *Global Environmental Change-Human and Policy Dimensions, 14*(2), 137–146.

Wang, X. N., Guo, M., Koppelaar, R., van Dam, K. H., Triantafyllidis, C. P., & Shah, N. (2018). A nexus approach for sustainable urban energy-water-waste systems planning and operation. *Environmental Science and Technology, 52*(5), 3257–3266.

Wiek, A., Binder, C., & Scholz, R. W. (2006). Functions of scenarios in transition processes. *Futures, 38*(7), 740–766.

Yang, L. S., & Lin, B. Q. (2016). Carbon dioxide-emission in China's power industry: Evidence and policy implications. *Renewable and Sustainable Energy Reviews, 60,* 258–267.

Chapter 7
Preliminary Exploration of Sustainability Solutions for Wastewater Management Services: A Case Study of Organic Carbon Regulation

7.1 Overview

In Chap. 3, the life cycle assessment (LCA) approach was used to perform an environmental impact inventory analysis of typical wastewater treatment processes. On this basis, assessment elements, such as the recovery potential of usable matter in wastewater and sludge, were incorporated for the construction of a novel assessment system for multi-objective management of municipal wastewater treatment in Chap. 4, and a method for the determination of an indicator weighting system was subsequently proposed in Chap. 5. In the following chapter, the time series prediction method was adopted to perform scenario analysis and predictions for the multi-objective management of municipal wastewater treatment in developed and developing countries. The analysis results indicate that the simultaneous bioenergy and nutrient recovery scenario will be of great significance to the achievement of sustained management of wastewater treatment in the future. In general, the objectives of all preceding chapters are as follows: (1) to determine the importance of the environmental footprint and resource recovery potential during the wastewater treatment process and address the current lack of assessment systems; (2) to contemplate the possibility of achieving energy conservation, carbon emission reduction, and resource recovery in current wastewater treatment processes and provide a theoretical basis and scientific foundation for research on regulation models for wastewater treatment.

In current wastewater treatment processes, the transfer and transformation are two basic processes used for the removal of pollutants. Physical treatment units, such as screening, sedimentation, clarification, and filtration, enable the transfer of pollutants from the aqueous phase to the solid phase through separation, while biological treatment units transform pollutants into harmless or usable matter through microbial degradation. During the literature review process, it was found that processes involving the transfer of pollutants have a low energy consumption and low greenhouse gas (GHG) emissions . Aerobic biological transformation processes have high energy

X. Wang, *Energy Consumption, Chemical Use and Carbon Footprints of Wastewater Treatment Alternatives*, Springer Theses,
https://doi.org/10.1007/978-981-13-5983-5_7

consumption and produce the greatest amount of excess sludge and GHG emissions, while anaerobic biological transformation processes have low energy consumption and can transform organic matter into energy sources. Therefore, the modification and regulation of aerobic biological treatment processes are critical to achieving the reduction of the environmental footprint (e.g., energy consumption and GHG emissions) during the wastewater treatment process, while the regulation of anaerobic processes plays a crucial role in the resource recovery of pollutants.

Biodegradable organic matter in municipal wastewater is mainly composed of readily biodegradable organic carbon and slowly degradable organic carbon. Under anoxic conditions, these types of organic matter can be converted to short-chain fatty acids (SCFAs) by anaerobic bacteria (Wiesmann et al. 2007). Based on previous studies, the most common SCFAs found in wastewater treatment plants (WWTPs) are acetate and propionate (Ahn and Speece 2006; Thomas et al. 2003; Zeng et al. 2006). Theoretically speaking, because different organic compounds have different molecular structures, the energy requirements for the degradation also differ. Organic compounds with complex molecular structures = consume more energy during the degradation process (Talaro 2008); however, from the perspective of energy storage within organic compounds, macromolecular organic compounds have a higher energy recovery potential (Shizas and Bagley 2004). Although such analyses do not consider actual reaction systems and biochemical kinetic processes, the analysis of the influence of different organic carbon compounds on material and energy flows and the effective regulation of the targeted transfer and transformation of such compounds in the wastewater treatment process may form an effective strategy for the achievement of energy conservation, reduced carbon emissions, and resource recovery in wastewater treatment. It is widely known that organic carbon is the substrate for microbe growth; therefore, a certain concentration of organic carbon must be maintained to enable microbe growth and reproduction. However, differences in the metabolic pathways of different microbes pose certain difficulties for research, which to a certain degree explains the current lack of relevant studies. Nevertheless, some researchers asserted that short-chain organic substrates, such as SCFAs, can be rapidly metabolized by phosphorus-accumulating organisms (PAO) to form polyhydroxyalkanoates (PHAs), thereby promoting the transfer of soluble orthophosphate (SOP) from the wastewater to the solid phase (sludge) (Jiang et al. 2009; Tong and Chen 2007; Zhang et al. 2010). This provides certain inspirations for the regulation of wastewater treatment: During the wastewater treatment process, the degradation routes of organic carbon may be regulated such that more organic carbon is stored in the form of PHAs instead of being aerobically degraded. Concurrently, the transfer of phosphorus in wastewater to excess sludge may enable the achievement of phosphorus recovery in subsequent processes. At present, the quantification and regulation models of such a concept, including the influence of organic carbon sources and relevant metabolic pathways, have not been studied.

In this study, organic carbon compounds commonly present in wastewater (rapidly degradable organic matter, acetate, and propionate) were used as study subjects and whole-process simulation technology was employed for the analysis of the influence of organic carbon on the energy consumption and resource recovery potential in wastewater treatment. Subsequently, the transfer and transformation of organic carbon were used as regulatory means to preliminarily study the targeted regulation of multi-objective management of wastewater treatment.

7.2 Experimental Design

7.2.1 Study Methods

As stated in the literature review section, many domestic and foreign studies have been conducted on energy conservation and GHG emissions in wastewater treatment processes. However, the studies have been limited to the experimental exploration of certain aspects. In addition, due to constraints of the experimental conditions (e.g., manpower and time), it is difficult to conduct systematic research on energy conservation, reduced carbon emissions, and the resource recovery potential in wastewater treatment processes. Therefore, there is a lack of in-depth studies on the transfer and transformation mechanisms of pollutants and regulation methods in wastewater treatment. In view of these issues, the present study employs computational simulation technologies for the whole-process modeling and simulation of the wastewater treatment process on a numerical simulation platform to analyze the transfer and transformation mechanisms of pollutants from a biochemical kinetic perspective and to investigate the influence of the type of organic carbon compounds on the energy consumption and resource recovery potential of pollutants during wastewater treatment. Finally, the numerical simulation platform was utilized for the preliminary study of a targeted regulation model for wastewater treatment.

7.2.2 Experimental Scheme

7.2.2.1 Influence of Organic Carbon on the Energy Consumption and Resource Recovery Potential of Pollutants During Wastewater Treatment

Biological Treatment System

The processing capacity of the WWTP and average influent temperature were set to 100,000 m^3/d and 20 °C, respectively, and the wastewater treatment process consisted of water and sludge treatment lines. To simulate conventional activated sludge systems, the water treatment line is mainly comprised of an aerator and a clarifier. The process flow diagram is shown in Fig. 7.1. The carbon sources in the influent of the three wastewater treatment systems F-ASP, P-ASP, and A-ASP were assumed to be rapidly degradable organic matter, acetate, and propionate, respectively, while all

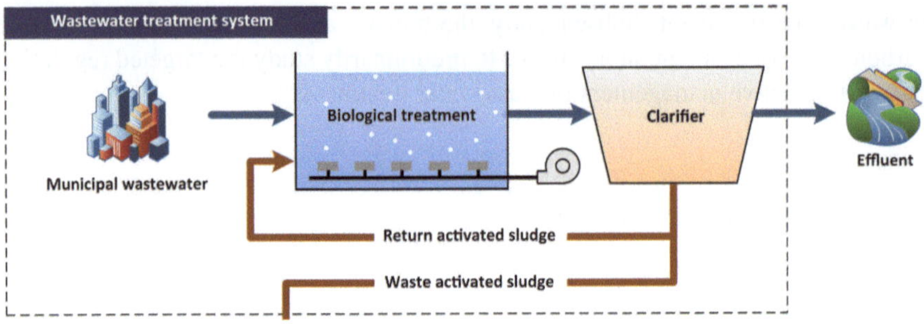

Fig. 7.1 Aerobic wastewater treatment process

other influent conditions were identical. The water quality parameters used in this study are described in the explanatory notes of Table 7.1. The operating volume of the aerator was set to 0.33×10^5 m^3, aeration holes were assumed to be uniformly distributed on the aerator bottom, the percentage of the open area was set to 10%, and the dissolved oxygen (DO) concentration was set to 2 mg/L. Based on the characteristics of conventional activated sludge treatment (Metcalf and Eddy 2003), solid residence times (SRTs) of 9, 11, and 15 days were selected for comparison in the present study. To investigate the influence of organic carbon on the resource recovery potential of the pollutants, the simultaneous bioenergy and nutrient recovery scenario constructed in Chap. 6 was utilized as follows: The excess sludge produced from the wastewater treatment system enters the anaerobic digester after thickening and the methane gas (CH_4) produced during the anaerobic digestion process is collected for bioenergy recovery. Simultaneously, nitrogen and phosphorus in the supernatant produced from the anaerobic digestion of sludge are recovered in the form of struvite, while the anaerobically digested sludge is dewatered and subsequently transported out of the plant for agricultural use via sludge composting.

Biochemical Kinetic Processes

Conventional aerobic biological wastewater treatment processes involve organic matter removal, nitrification, and dephosphorization. Therefore, ordinary heterotrophic organisms (OHOs), ammonia-oxidizing bacteria (AOB), nitrite-oxidizing bacteria (NOB), and phosphorus-accumulating organisms (PAOs) are functional bacteria commonly found in conventional activated sludge treatment systems (Metcalf and Eddy 2003). The mathematical models used in the present study adequately describe the aforementioned biochemical kinetic processes, which at least include the following processes: hydrolysis of slowly degradable organic matter; growth of OHOs through the utilization of rapidly degradable organic matter, acetate, and propionate; decline of OHOs; storage of acetate or propionate in the form of PHAs; aerobic growth of PAOs through the utilization of PHAs as substrates; decline of PAOs; aerobic growth

Table 7.1 Removal efficiencies in the three ASPs

Removal efficiency (%)[a]	SRT = 9 d			SRT = 11 d			SRT = 15 d		
	F-ASP	P-ASP	A-ASP	F-ASP	P-ASP	A-ASP	F-ASP	P-ASP	A-ASP
TP	38.0	40.9	90.4	34.9	72.7	90.0	30.9	85.1	88.9
SOP	41.1	44.3	97.1	38.1	79.3	97.0	34.3	94.0	97.0
NH_4^+–N	99.0	99.0	99.0	99.1	99.1	99.1	99.2	99.2	99.2
TN	25.7	25.4	26.4	24.1	23.2	25.3	22.0	20.0	23.8
COD	97.1	97.1	95.6	97.1	96.9	95.1	97.0	96.3	94.1
SS	94.8	94.9	94.7	94.7	94.6	94.4	94.4	94.3	93.8

Adapted from Wang et al. (2013), with permission from Elsevier, Copyright (2013)

[a] During model simulation, the following water quality parameters were specified for the influent: COD = 400 mg/L, TP = 7 mg P/L, SOP = 7 mg P/L, $NH4^+$-N = 40 mg N/L, TN = 40 mg N/L, SS = 171.4 mg/L.

of AOB and NOB; and decline of AOB and NOB. To achieve a comprehensive simulation of the aforementioned biochemical kinetic reactions, BioWin software (V.3.0., BW3-1952) was employed to construct and couple the relevant kinetic processes. The model characteristics of various functional bacteria and their kinetic behavior are identical to those described in the modeling process in Chap. 3; however, different influent conditions and process designs were used. The model descriptions of the anaerobic digestion of sludge are also provided in Chap. 3.

Experiment on the Metabolic Activity of OHOs

To investigate the influence of the metabolic activity of OHOs on the microbe growth, organic carbon transformation, and energy consumption, three biochemical kinetic and stoichiometric parameters that are representative for the metabolic activity of OHOs (maximum specific growth rate $\mu_{o,max}$, mg COD/mg COD d^{-1}; substrate half-saturation constant for growth $K_{s,o}$, mg COD/L; and yield coefficient $Y_{o,f}$, mg COD/mg COD) were selected for the experiment. In the computational simulation experiment, the metabolic activity of OHOs was enhanced by regulating the OHO model parameters. The metabolic activity increased as $\mu_{o,max}$ and $Y_{o,f}$ increase and $K_{s,o}$ decreases. The following values were used for the respective parameters: $\mu_{o,max}$: 1, 2, 3, 4, 5, and 6 mg COD/mg COD d^{-1}; $Y_{o,f}$: 0.2, 0.3, 0.4, 0.5, 0.6, and 0.7 mg COD/mg COD; and $K_{s,o}$: 2.0, 2.5, 3.0, 3.5, 4.0, 4.5, and 5.0 mg/L. When the value of $\mu_{o,max}$ was varied, $Y_{o,f}$ and $K_{s,o}$ were kept constant at 0.67 mg COD/mg COD and 5.0 mg/L, respectively; when the value of $Y_{o,f}$ was varied, $\mu_{o,max}$ and $K_{s,o}$ were kept constant at 3.2 mg COD/mg COD d^{-1} and 5.0 mg/L, respectively; when the value of $K_{s,o}$ was varied, $\mu_{o,max}$ and $Y_{o,f}$ were kept constant at 3.2 mg COD/mg COD d^{-1} and 0.67 mg COD/mg COD, respectively. Relevant information on wastewater treatment models published by the International Water Association (IWA) was used as reference for the selection of the parameter values (Makinia 2010).

Single-Substrate PHA Synthesis

To study the effect of different types of organic carbon on the PHA synthesis and avoid duplicate experimentation, supporting data were obtained by performing data mining using the experimental data reported by Dr. Bo Qu of our research group in a previous study. The raw data mainly originate from Sect. 5.2.2 of the doctoral dissertation of Dr. Qu entitled "Production of biodegradable thermoplastics—PHA from excess sludge." The relevant experimental methods and detailed data are provided in the aforementioned doctoral dissertation (Qu 2008).

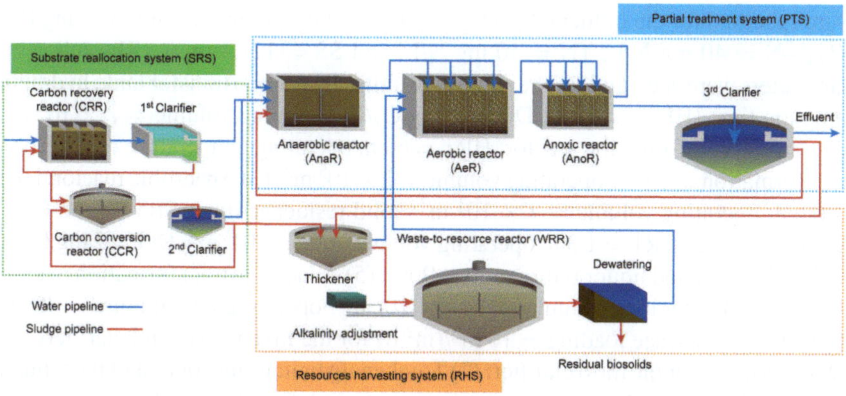

Fig. 7.2 Model for the targeted regulation of wastewater treatment aimed at multi-objective management. Adapted from Wang et al. (2018)

7.2.2.2 Preliminary Study of the Targeted Regulation of Wastewater Treatment

To achieve the objectives of energy conservation, reduced carbon emissions, and resource recovery, a model for the targeted regulation of wastewater treatment has been preliminary proposed based on the research conducted in the preceding chapters, as shown in Fig. 7.2.

In the targeted regulation model, the influent wastewater of the municipal WWTP first enters the carbon transfer tank and undergoes a biologically enhanced transfer process, which involves the rapid transfer of organic matter from wastewater to sludge within a short period of time. Subsequently, the resultant water/sludge mixture enters the first clarifier and is separated into water and sludge. After the separation, the wastewater enters the subsequent wastewater treatment units, while the sludge enters the carbon transformation reactor, where most of the organic matter in the sludge is converted to acetate. The wastewater produced in the second clarifier also enters the wastewater treatment units, while the sludge from the second clarifier enters the thickening unit together with excess sludge produced in the wastewater treatment units. After thickening, the sludge enters the anaerobic transformation reactor, where CH_4 is produced from anaerobic digestion of sludge and struvite is produced from the supernatant of the digested sludge. Lastly, the anaerobically treated sludge is dewatered and transported out of the WWTP for agricultural use. The wastewater treatment units primarily comprise the anaerobic and aerobic reactors. Effluents from the first and second clarifiers enter the anaerobic reactor, while the effluent from the dewatering unit enters the aerobic reactor. Additionally, a portion of the mixed liquor from the aerobic reactor is recycled back to the anaerobic reactor.

The study in this chapter was conducted using a computer simulation platform. In the model of the targeted regulation of wastewater treatment, the processing capacity of the WWTP was set to 100,000 m^3/d and the following water quality parameters were specified for the influent: COD = 400 mg/L, TP = 7 mg P/L, SOP = 7 mg P/L, NH_4^+–N = 40 mg N/L, TN = 40 mg N/L, and SS = 171.4 mg/L. The following parameters were specified for the respective units: (1) Carbon transfer tank: hydraulic retention time (HRT) = 0.5 h, DO = 2 mg O_2/L, operating volume = 2×10^3 m^3; (2) Carbon transformation reactor: HRT = 60 h, operating temperature of anaerobic transformation = 20 °C, operating volume = 2×10^5 m^3; (3) Anaerobic reactor: HRT = 0.6 h, operating volume = 7×10^3 m^3; solid residence time (SRT) = 3.5 d; (4) Aerobic reactor: HRT = 1.3 h, operating volume = 1.5×10^4 m^3, DO = 2 mg O_2/L, SRT = 3.5 d, mixed liquor return ratio = 90%; (5) Anaerobic transformation reactor: HRT = 30 d, operating volume = 1.14×10^4 m^3; operating temperature = 35 °C; (6) Clarifiers: surface loading = 16 $m^3/(m^2$ d) for the first and second clarifiers and 19.4 $m^3/(m^2$ d) for the third clarifier; (7) Thickening unit: water content of the effluent sludge ≈ 95%; and (8) Dewatering unit: water content of effluent sludge ≈ 85%.

7.2.3 Calculation Methods

The estimation models and calculation methods used in this chapter for the determination of energy consumption, GHG emissions, and the various resource recovery potentials are identical to those described in Sect. 4.5.3.

7.3 Influence of Organic Carbon on Energy Consumption and Resource Recovery Potential of Pollutants

7.3.1 Analysis of Removal Efficiencies in Wastewater Treatment

The removal efficiencies for total phosphorus (TP), soluble orthophosphate (SOP), ammonia nitrogen (NH_4^+–N), total nitrogen (TN), chemical oxygen demand (COD), and suspended solids (SS) in the various wastewater treatment processes (F-ASP, P-ASP, and A-ASP) with SRTs of 9, 11, and 15 d are shown in Table 7.1. The removal efficiencies for TP and SOP are higher when the substrates in the influent wastewater are SCFAs, with the increase being particularly pronounced for acetate. This result is similar to the phenomenon described in the previous literature, which stated that the reactors exhibit an enhanced phosphorus removal capability when acetate or propionate is used as the substrates (Guisasola et al. 2004; Pijuan et al. 2004).

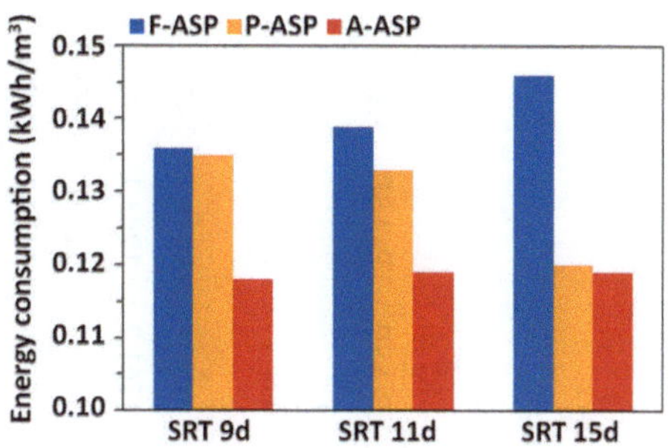

Fig. 7.3 Energy consumption of substrate oxidation in the three ASPs. Adapted from Wang et al. (2013), with permission from Elsevier, Copyright (2013)

7.3.2 Effect of Different Organic Carbon Sources on the Energy Consumption

In this study, the aeration energy consumption during the nitrification process remained relatively stable. Therefore, the effect of different types of organic carbon compounds on the energy consumption by nitrification is not discussed here. Figure 7.3 shows the energy consumption of organic carbon oxidation in the aerator.

Figure 7.3 shows that the type of organic carbon source has a certain effect on the energy consumption in the aerobic wastewater treatment process. Among the three wastewater treatment processes, F-ASP, which utilizes rapidly degradable organic matter as the carbon source, requires a higher energy input during the aerobic degradation of organic carbon; by contrast, A-ASP, which utilizes acetate as the carbon source, has the lowest energy consumption during organic carbon degradation. Table 7.2 summarizes the aerobic transformation processes of organic carbon, SOP, and PHAs in the three ASPs. When the SRT = 15 d, the rate of carbon oxidation during F-ASP is significantly higher than that of P-ASP and A-ASP (1761.2 vs. 1374.1 and 1286.1 mg COD/L d). This shows that the F-ASP has a higher specific energy consumption during the organic carbon degradation process (0.146 vs. 0.120 and 0.119 kWh/m^3) and a greater amount of organic carbon is oxidatively degraded to CO_2 (31040.5 vs. 27212.6 and 26678.4 m^3/d). In addition, compared with P-ASP or A-ASP, F-ASP has an extremely low PAO growth rate (0 vs. 111.0 and 369.6 mg COD/L d) and an extremely low rate of PHA synthesis from organic carbon (0 vs. 350.3 and 665.7 mg COD/L d). In other words, organic carbon is mainly oxidatively degraded to CO_2 and H_2O by OHOs during F-ASP, which leads to the relatively high organic carbon oxidation rate and energy consumption during F-ASP. However, substrate competition exists between OHOs and PAOs during P-ASP and A-ASP, where the bulk of the carbon substrate is controlled by PAOs and stored in the form of PHAs, thereby leading to lower carbon oxidation rates and energy consumption levels of these two treatment systems.

Table 7.2 also shows that the PHA synthesis and degradation rates during A-ASP are both higher than that during P-ASP (synthesis rate: 665.7 vs. 350.3 mg COD/L d; degradation rate: 369.6 vs. 111.0 mg COD/L d). This shows that acetate produces a higher amount of PHAs compared with an equivalent amount of pro-

Table 7.2 Aerobic transformations of organic carbon, SOP, and PHAs in the three ASPs[a]

Item	SRT = 9 d			SRT = 11 d			SRT = 15 d		
	F-ASP	P-ASP	A-ASP	F-ASP	P-ASP	A-ASP	F-ASP	P-ASP	A-ASP
Initial SOP concentration (mg P/L)	7	7	7	7	7	7	7	7	7
SOP concentration at the end of the aerobic transformation (mg P/L)	4.1	3.9	0.2	4.3	1.5	0.2	4.6	0.4	0.2
SOP removal rate (%)	41.1	44.3	97.1	38.1	79.3	97.0	34.3	94.0	97.0
Carbon source oxidation rate of OHOs (mg COD/L d)	1627.9	1548.0	980.8	1685.3	1476.1	960.7	1761.2	1263.1	916.5
Carbon source oxidation rate of PAOs (mg COD/L d)	0	3.4	270.7	0	37.4	308.8	0	111.0	369.6
Carbon source oxidation rate (mg COD/L d)	1627.9	1551.4	1251.5	1685.3	1513.5	1269.5	1761.2	1374.1	1286.1
CO_2 production (m^3/d)	29306.5	29015.4	26338.0	29889.0	28890.9	26542.0	31040.5	27212.6	26678.4
Synthesis rate of PHAs (mg COD/L d)	0	10.8	491.2	0	118.6	558.3	0	350.3	665.7
Difference between synthesis and degradation rates of PHAs[b] (mg COD/L d)	0	7.4	220.5	0	81.2	249.5	0	239.3	296.1
Difference between synthesis rate of PHAs and carbon source oxidation rate (–)	0	0.01	0.39	0	0.08	0.44	0	0.25	0.52
PHA concentration at the end of the aerobic transformation (mg COD/L)	0	0.5	47.4	0	5.5	51.4	0	16.6	58.3

[a]All data were obtained from numerical simulations; [b]The degradation rate of the PHAs is equivalent to the carbon source oxidation rate of the PAOs
Adapted from Wang et al. (2013), with permission from Elsevier, Copyright (2013)

Table 7.3 Kinetic and stoichiometric parameters and composition of polymers in single-substrate experiments

Carbon substrate	q_{SOUR}	$-q_S$	$-q_N$	q_{HB}	q_{HV}	Y_X	Y_{HB}	Y_{HV}	Type of PHA
Acetate	14.72	0.35	0.009	0.13	–	0.13	0.39	–	HB
Propionate	3.67	0.05	0.002	–	0.014	0.18	–	0.27	HV

q_{SOUR}: 10^{-2} C mmol/C mmol h; qS, q_{HB}, q_{HV}: C mmol/C mmol h; Y_X, Y_{HB}, Y_{HV}: C mmol/C mmol h; q_N: N mmol/C mmol h

Adapted from Wang et al. (2013), with permission from Elsevier, Copyright (2013)

pionate. To further elucidate the similarities and differences of the organic carbon storage between A-ASP and P-ASP, data mining and consolidation were performed using experimental data obtained by Dr. Bo Qu in a previous study. The results are shown in Table 7.3 and Fig. 7.4. In Table 7.3, q_{SOUR} is the specific oxygen uptake rate, q_S is the specific substrate uptake rate, q_N is the specific ammonia nitrogen uptake rate, q_{HV} is the specific hydroxyvalerate (HV) storage rate, q_{HB} is the specific hydroxybutyrate (HB) storage rate, Y_{HV} is the HV yield, Y_{HB} is the HB yield, and Y_X is the active biomass yield. The specific uptake rate of acetate is much higher than that of propionate (0.35 vs. 0.05 C mmol/C mmol h), which shows that a greater amount of polymer is synthesized when acetate is used as the substrate (0.39 vs. 0.27 mmol/C mmol h). Based on a previous study by Wang et al., acetate is first converted into acetyl-coenzyme A (acetyl-CoA) by microbes when it is used as the substrate. A portion of the acetyl-CoA subsequently enters the tricarboxylic acid cycle, thereby reducing the energy required for polymer synthesis (Wang et al. 2009). However, when propionate is used as the substrate, propionyl-coenzyme A (propionyl-CoA) will react with acetyl-CoA to form hydroxyvaleryl-CoA. The acetyl-CoA involved in the reaction is formed from the decarboxylation of propionyl-CoA, which is regarded as the most energy-consuming metabolic pathway during PHA synthesis (Doi et al. 1987). The PHA concentration at the end of A-ASP is therefore higher than that of P-ASP (58.3 vs. 16.6 mg COD/L). Lemos et al. also observed that the amount of synthesized PHAs decreases with an increase in the chain lengths of the SCFA substrates (Lemos et al. 1998).

Theoretically speaking, the oxidative degradation of simple organic compounds requires less energy due to the short-chain structures of these compounds. However, in a complex biochemical reaction system, biochemical kinetic processes have a more profound effect on the energy consumption. In contrast to F-ASP, the organic carbon in wastewater is not completely oxidatively degraded to CO_2 and H_2O by OHOs in A-ASP; instead, it is stored in polymer form, that is, PHAs. In other words, the competition for organic carbon between OHOs and PAOs may be instrumental in determining the level of energy consumption by aerobic processes. For this reason, we studied the influence of the metabolic activity of OHOs on the microbe growth, organic carbon storage, and energy consumption of aerobic transformation. The results are shown in Fig. 7.5. Clearly, if the metabolic activity of the OHOs can be

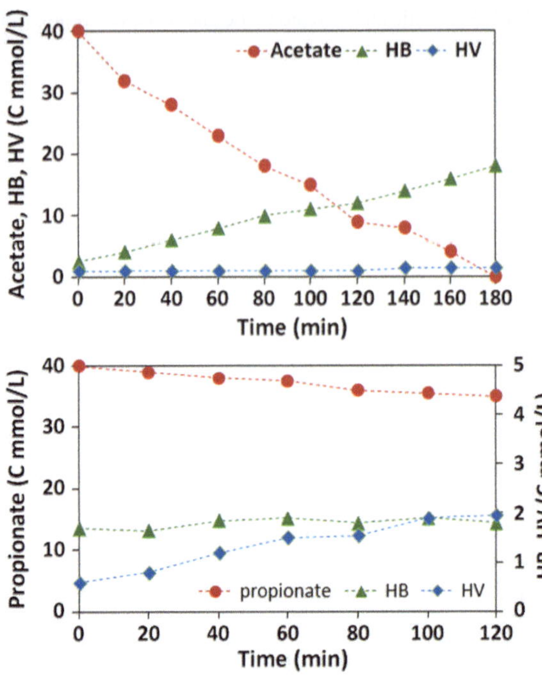

Fig. 7.4 Profiles for substrate uptake and storage in polymers in single-substrate experiments. Adapted from Wang et al. (2013), with permission from Elsevier, Copyright (2013)

decreased, thereby reducing its competitiveness for organic carbon metabolism, more organic carbon is stored in the form of PHAs, even with the occurrence of oxidative degradation of PHAs by PAOs. In addition, Table 7.2 shows that the ratios of the PHA synthesis rate to organic carbon oxidation rate for F-ASP, P-ASP, and A-ASP are 0, 0.25, and 0.52, respectively. This indicates that only 48% of the organic carbon in A-ASP is degraded through the aerobic growth of microbes, while the remaining 52% of organic carbon is stored in the form of PHAs. Therefore, the energy consumption during wastewater treatment can be effectively reduced by using the A-ASP system.

7.3.3 *Influence of Different Organic Carbon Sources on the Resource Recovery Potential*

The data shown in Table 7.4 indicate the resource recovery potentials of excess sludge produced by various wastewater treatment systems. Because the data trends for different SRTs are largely similar, only the trends for an SRT = 15 d are discussed below. In the simultaneous bioenergy and nutrient recovery scenario, the excess sludge produced from F-ASP has a higher specific bioenergy recovery potential than that of P-ASP and A-ASP (0.141 vs. 0.114 and 0.126 kWh/m^3). In addition, the energy consumption required for anaerobic sludge digestion is lower during F-ASP

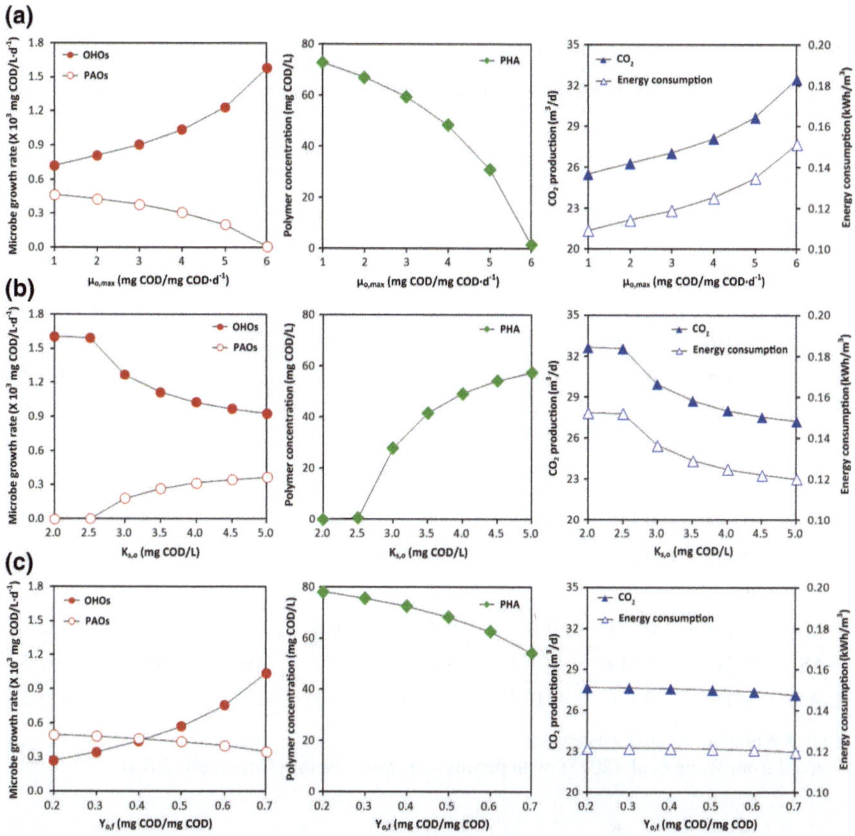

Fig. 7.5 Influence of the metabolic activity of OHOs on the microbe growth, carbon storage, and energy consumption of oxidation

than during P-ASP and A-ASP (0.038 vs. 0.041 and 0.049 kWh/m^3). Therefore, the excess sludge generated from F-ASP has a higher net bioenergy recovery (0.103 vs. 0.073 and 0.077 kWh/m^3). Regarding the recovery of other resources, the struvite recovery potential of F-ASP is lower than that of P-ASP and A-ASP (123.4 vs. 4240.8 and 2334.8 kg ISS/d). In the nutrient recovery scenario, A-ASP has the highest N and P recovery potential among all three systems (N: 866.1 vs. 741 and 687.5 kg N/d; P: 623.3 vs. 222.2 and 598.4 kg P/d).

Table 7.5 shows the concentrations of volatile solids (VS), inorganic solids (IS), and total solids (TS) in the excess sludge produced from various wastewater treatment systems. Compared with the excess sludge produced from F-ASP and P-ASP, the excess sludge from A-ASP has relatively high concentrations of volatile active suspended solids (VASS) and PHAs (VASS: 5073.63 vs. 3777.04 and 3665.96 mg VS/L; PHAs: 72.34 vs. 0 and 20.59 mg VS/L). However, the excess sludge from F-ASP has higher concentrations of endogenous respiration products and slowly degrad-

Table 7.4 Resource recovery potential of the excess sludge in the three ASPs with the following SRTs: (a) 9 days; (b) 11 days; (c) 15 days

System	Simultaneous bioenergy and nutrient recovery scenario						Nutrient recovery scenario	
	Bioenergy recovery[a]			Recovery of other resources				
	Energy consumption	Energy production	Net recovery	Struvite (kg ISS/d)	N (kg N/d)	P (kg P/d)	N (kg N/d)	P (kg P/d)
a								
F-ASP	0.046	0.206	0.160	129.5	693.4	264.0	932.3	274.4
P-ASP	0.045	0.196	0.151	448.9	663.9	275.8	896.6	294.4
A-ASP	0.054	0.180	0.127	2061.7	793.2	567.8	991.7	635.5
b								
F-ASP	0.043	0.178	0.135	127.3	648.4	243.0	849.7	252.0
P-ASP	0.044	0.159	0.115	2926.6	618.4	431.5	806.9	512.9
A-ASP	0.052	0.157	0.105	2163.5	770.3	562.6	937.2	631.4
c								
F-ASP	0.038	0.141	0.103	123.4	686.6	215.0	742.1	222.2
P-ASP	0.041	0.114	0.073	4240.8	543.3	491.5	687.5	598.4
A-ASP	0.049	0.126	0.077	2334.8	736.3	552.7	866.1	623.3

[a]Unit: kWh/m^3 of treated wastewater
Adapted from Wang et al. (2013), with permission from Elsevier, Copyright (2013)

able organic matter than that of P-ASP and A-ASP (2453.84 + 73.44 = 2527.28 vs. 2085.51 + 48.30 = 2133.81 and 2346.30 + 31.35 = 2377.65 mg VS/L). The polyphosphate (poly-P) and organic ash content of the excess sludge produced from A-ASP are also higher than that of F-ASP and P-ASP (612.64 + 1112.35 = 1724.99 vs. 0 + 542.19 = 542.19 and 617.48 + 1007.42 = 1624.90 mg IS/L). Therefore, the VS and IS concentrations of the excess sludge from A-ASP are higher than that of F-ASP and P-ASP (VS: 7523.61 vs. 6304.32 and 5820.37 mg VS/L; IS: 2974.89 vs. 1773.84 and 2862.52 mg IS/L), which results in the highest VS concentration (10498.50 vs. 8078.16 and 8682.89 mg TS/L). The VS/TS ratios of the excess sludge from F-ASP, P-ASP, and A-ASP are 0.78, 0.67, and 0.72, respectively. Kim and Sandor et al. found that higher VS/TS ratios of excess sludge are associated with a higher methanation efficiency and gas production during the anaerobic sludge digestion process (Kim et al. 2003; Sandor et al. 2011). Although organic carbon is stored and transferred to sludge during A-ASP instead of being oxidatively degraded, the degree of organic carbon methanation largely depends on the proportion of organic carbon. In other words, in addition to the content of volatile organic components, the VS/TS ratio of the excess sludge is the key factor that influences the methanation of organic carbon.

Table 7.5 VS, IS, and TS concentrations and VS/TS ratio of raw excess sludge in the three ASPs[a]

Item	F-ASP			P-ASP			A-ASP		
	9 d	11 d	15 d	9 d	11 d	15 d	9 d	11 d	15 d
PHAs	0	0	0	0.62	6.79	20.59	58.56	63.68	72.34
Endogenous respiration products	1289.22	1685.46	2453.84	1228.65	1550.37	2085.51	1150.61	1542.28	2346.30
Slowly degradable organic matter	63.05	67.53	73.44	59.83	57.49	48.30	32.47	32.33	31.35
VASS[b]	3244.01	3473.08	3777.04	3112.48	3328.64	3665.96	3742.86	4247.70	5073.63
VS[c]	4596.28	5226.07	6304.32	4401.57	4943.29	5820.37	4984.50	5885.99	7523.61
ISS	748.43	916.66	1231.65	747.10	919.02	1237.62	755.57	927.77	1249.90
Poly-P	0	0	0	28.80	242.90	617.48	340.03	433.35	612.64
Organic ash	394.46	448.88	542.19	401.75	584.82	1007.42	689.97	837.87	1112.35
IS[d]	1142.89	1365.54	1773.84	1177.65	1746.74	2862.52	1785.57	2198.99	2974.89
TS[e]	5739.17	6591.61	8078.16	5579.22	6690.03	8682.89	6770.07	8084.98	10498.50
VS/TS ratio	0.80	0.79	0.78	0.79	0.74	0.67	0.74	0.73	0.72

[a]Unit: mg SS/L. [b]VASS concentration obtained by summing the products of various microbe concentrations (XOHO, XPAO, XAOB, and XNOB) and their respective COD/VS ratios. [c]VS concentration obtained by adding the VASS concentration to the sum of products of the various particulate organic matter concentrations and their respective COD/VS ratios. [d]IS includes inorganic components, total settled solids, poly-P, and organic ash. [e]TS consists of VS and IS
Adapted from Wang et al. (2013), with permission from Elsevier, Copyright (2013)

Table 7.6 Treatment efficiencies in the targeted regulation model

Effluent indicator	Concentration (mg/L)	Removal efficiency (%)
COD	28.0	93.0
BOD	2.7	98.6
NH_4^+–N	3.0	92.5
TN	14.8	63.0
TP	0.9	87.1
SOP	0.6	91.4
SS	3.0	98.3

7.4 Preliminary Study of the Targeted Regulation Model of Wastewater Treatment Aimed at Multi-objective Management

7.4.1 Analysis of the Treatment Efficiency

Firstly, the wastewater treatment efficiency in the targeted regulation model was investigated. The concentrations and removal efficiencies of TP, SOP, NH_4^+–N, TN, COD, and SS are shown in Table 7.6. The effluent in the targeted regulation model satisfies the water quality requirements of the Class 1A standard of the Pollutant Discharge Standard of Municipal Wastewater Treatment Plants (GB18918-2002).

7.4.2 Reactor Volume Analysis

Table 7.7 shows the volumes of the main reactors and tanks in the conventional wastewater treatment and targeted regulation models. Based on the process flow of wastewater treatment, the total reactor volume for the conventional model is 1.86×10^5 m^3, which accounts for 73.8% of the total volume, while the total reactor volume for the targeted regulation model is 0.24×10^5 m^3, which accounts for only 8.6% of the total volume. In other words, with the adoption of the targeted regulation model, an 80% reduction in the total volume of conventional wastewater treatment reactors is achieved. However, the total clarifier volume increases. With regard to the process flow of the sludge treatment, because the targeted regulation model involves the addition of a carbon transformation reactor, the total reactor volume of the sludge treatment process was 2.11×10^5 m^3, which accounts for 75.9% of the total volume and is almost 8 times the total reactor volume of the conventional model. The total volume of the main reactors and tanks of the conventional model is 2.52×10^5 m^3, while the corresponding value for the targeted regulation model is 2.78×10^5 m^3, which is 1.1 times that of the conventional model.

Table 7.7 Average reactor and tank volumes in the conventional wastewater treatment and targeted regulation models (unit: 10^3 m^3)

Unit	Conventional model	Targeted regulation model
Anaerobic reactor	30	7
Anoxic reactor	51	–
Aerobic reactor	105	15
Clarifier	40	45
Carbon transfer tank	–	2
Carbon transformation reactor	–	200
Anaerobic sludge digester	26	–
Anaerobic transformation reactor	–	11
Total volume	252	278

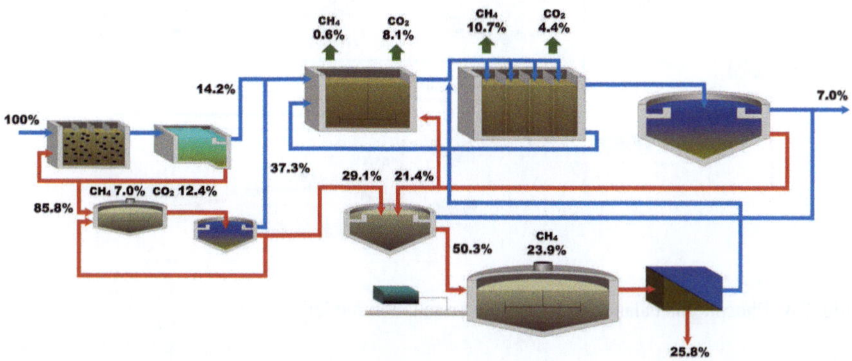

Fig. 7.6 Carbon balance in the targeted regulation model

7.4.3 *Material Balance Analysis*

To achieve a better understanding of the material transfer and transformation in the targeted regulation model of wastewater treatment, the material balances of carbon, nitrogen, and phosphorus during the wastewater and sludge treatment processes were determined. The results are shown in Figs. 7.6, 7.7, and 7.8.

As shown in Fig. 7.6, when wastewater flows through the carbon transfer tank, the organic carbon in the effluent is hardly oxidatively degraded. Instead, it is transferred to sludge within an extremely short residence time. After clarification, approximately 86% of the organic carbon, which is present in the sludge, enters the carbon transformation reactor for acidification. During anaerobic transformation, 7.0 and 12.4% of the organic carbon escape into the atmosphere as CH_4 and CO_2, respectively. When the effluent from the first clarifier enters the anaerobic reactor, 0.6 and 8.1% of the organic carbon in the wastewater also escape into the atmosphere as CH_4 and CO_2, respectively. The anaerobic reactor effluent then enters the subsequent

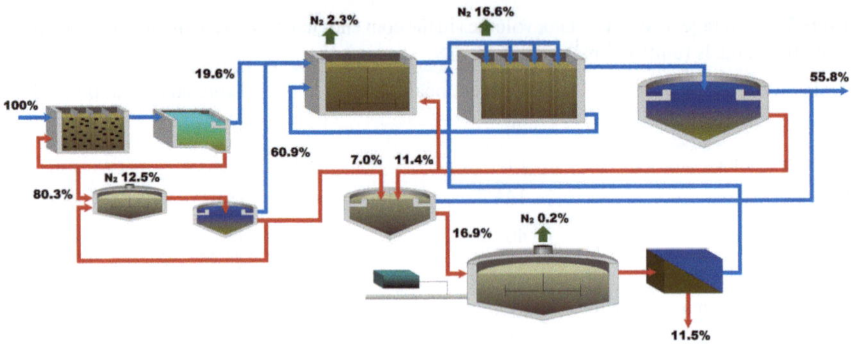

Fig. 7.7 Nitrogen balance in the targeted regulation model

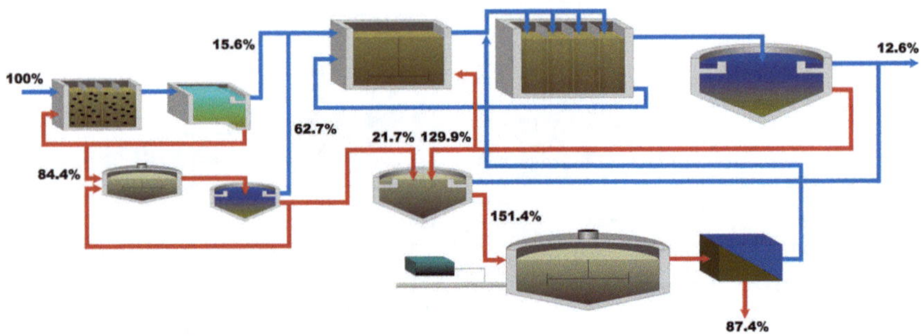

Fig. 7.8 Phosphorus balance in the targeted regulation model

aerobic reactor, where the soluble CH_4 in the wastewater is further blown off by aeration and released into the atmosphere (approximately 10.7% of the total carbon in the WWTP influent). In the aerobic unit, the organic carbon that is oxidatively degraded to CO_2 accounts for 4.4% of the total carbon in the WWTP influent. After thickening, all excess sludge enters the anaerobic transformation reactor for anaerobic digestion, which produces CH_4 (the recovered CH_4 accounts for 23.9% of the total carbon in the WWTP influent). The dewatered anaerobically digested sludge contains approximately 25.8% of the total carbon in the WWTP influent, while the treated wastewater only contains approximately 6.8% of the total carbon. In the targeted regulation model of the wastewater treatment, 24.9% of the total carbon in the WWTP influent is oxidatively degraded to CO_2. In addition, under the premise that the WWTP effluent meets the relevant water quality standard, 56.7% of the carbon in the WWTP influent can be recovered through resource recovery routes.

Figure 7.7 shows that approximately 11.5% of the nitrogen can be recovered in the targeted regulation model (1.5% is recovered in the form of struvite, while the remaining 10% of N is recovered through sludge composting). As shown in Fig. 7.8, given that the effluent meets the relevant quality standard, approximately 27.8% of

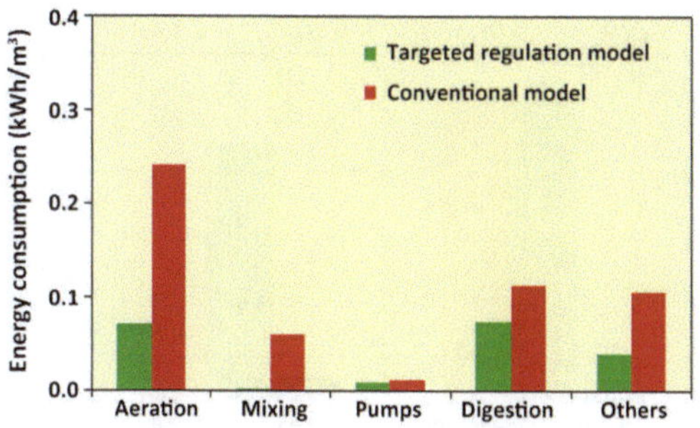

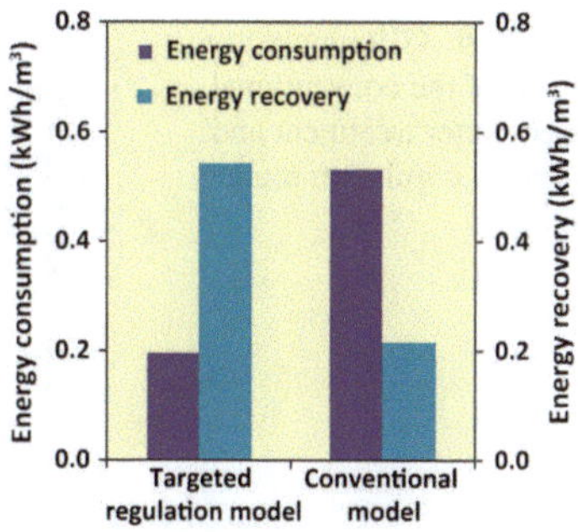

Fig. 7.9 Energy consumption and recovery levels of the conventional wastewater treatment and targeted regulation models

the total phosphorus in the wastewater can be recovered in the form of struvite, while 59.6% can be recovered through sludge composting for agricultural use.

7.4.4 Energy Balance Analysis

Using the calculation methods stated in Sect. 7.2.3, the energy consumption levels of the main processes in the targeted regulation model were calculated. Subsequently, these energy consumption values were compared with the energy consumption levels of the conventional wastewater treatment model (obtained from the average energy consumption values of six conventional wastewater treatment systems in Chap. 6). On this basis, the energy recovery potentials of both models (using data obtained from Chap. 6) were analyzed. The results are shown in Fig. 7.9.

In the targeted regulation model of wastewater treatment, the energy consumed by aeration is 0.072 kWh/m^3, which accounts for only 36.7% of the total energy consumption; this includes the energy consumed by the aeration units of the carbon transfer tank and aerobic reactor. The aeration energy consumption of the conventional wastewater treatment model is only contributed by the aerobic unit of the wastewater treatment system. The average energy consumption is 0.240 kWh/m^3, which is approximately 45.2% of the total energy consumption and 3.3 times that of the energy consumed by aeration in the targeted regulation model. Therefore, the adoption of the targeted regulation model can lead to a reduction of the aeration energy consumption by approximately 70% compared with the conventional model. The total energy consumption for mixing in the conventional model, mainly contributed by the anaerobic reactor, anoxic reactor, and anaerobic sludge digester, is estimated to be 0.06 kWh/m^3, which accounts for 11.4% of the total energy consumption. In the targeted regulation model, the total energy consumption for mixing is mainly contributed by the carbon transfer tank, anaerobic reactor, and anaerobic transformation reactor. However, it accounts for only 0.01% of the total energy

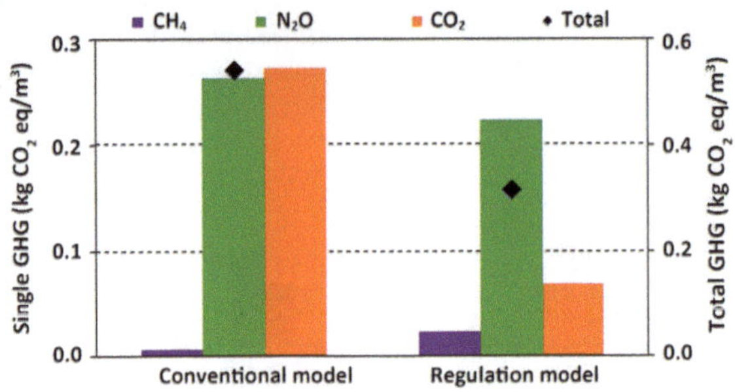

Fig. 7.10 GHG emission levels of the conventional wastewater treatment and targeted regulation models

consumption of the model. In the conventional and targeted regulation models, the combined energy consumption of the mixed liquor recirculation pump and return activated sludge pump are 0.012 and 0.010 kWh/m^3, respectively, which is not significantly different. During the anaerobic sludge digestion process, the energy consumption of the targeted regulation model is 0.075 kWh/m^3, which accounts for 38.3% of the total energy consumption, while the corresponding energy consumption of the conventional model is 0.11 kWh/m^3, which accounts for 21.2% of the total energy consumption.

With regard to the total energy consumption and given that the effluent satisfies the water quality requirements of the Class 1A standard, the average energy consumption of the conventional model is 0.53 kWh/m^3, while the corresponding value for the targeted regulation model is 0.20 kWh/m^3, which is only 36.8% of the energy consumed in the conventional model. Regarding the energy recovery, for every cubic meter of wastewater treated using the targeted regulation model, 0.54 kWh of energy can be recovered in the form of CH_4 produced from anaerobic sludge digestion. In contrast, the corresponding energy recovery potential of the conventional model is only 0.22 kWh/m^3. Therefore, from the energy balance perspective, for every cubic meter of wastewater treated, an energy input of 0.31 kWh is required in the conventional model. However, a net energy recovery of 0.34 kWh can be achieved using the targeted regulation model constructed in the present chapter.

7.4.5 Analysis of GHG Emissions

Using the estimation models and calculations described in Sect. 7.2.3, the GHG emission levels and contributions by the respective GHGs to the targeted regulation model were determined. These values were compared with the corresponding data for the conventional wastewater treatment model (obtained from Chap. 6). The results are shown in Fig. 7.10.

In the targeted regulation model, the CO_2-equivalent emissions of methane (CH_4), nitrous oxide (N_2O), and carbon dioxide (CO_2) are 0.023, 0.225, and 0.069 kg CO_2-e/m^3, respectively. The corresponding emissions in the conventional model are 0.007,

0.264, and 0.273 kg CO_2-e/m^3, respectively. Compared with the conventional model, the targeted regulation model has advantages with respect to the reduced N_2O and CO_2 emissions, with the respective reductions being 14.9 and 74.6%, respectively. The reduction in the N_2O emissions is primarily attributed to the recovery of ammonia nitrogen during struvite synthesis, which consequently reduces the N_2O produced during the denitrification process. The main reason for the CO_2 emission reduction is the energy self-sufficiency that can be achieved with the targeted regulation model. Compared with the power generation using traditional fossil fuels, the amount of CO_2 produced through CH_4 combustion is lower, thereby favoring the reduction of the CO_2 emissions. However, the targeted regulation model does not have an advantage over the conventional model in terms of the CH_4 emission reduction because the former produces tailwater and excess sludge with higher concentrations of soluble CH_4, thereby leading to the release of a higher amount of CH_4 from wastewater or sludge. In general, the total GHG emissions in the targeted regulation and conventional models are 0.312 and 0.545 kg CO_2-e/m^3, respectively. Therefore, compared with the conventional model, the GHG emissions can be reduced by 41.7% by adopting the targeted regulation model.

7.5 Summary

(1) The type of organic carbon present in wastewater significantly affects the energy consumption, GHG emissions, and resource recovery potential of pollutants during the wastewater treatment process. Organic carbon in the form of SCFAs, especially acetate, favors the storage of organic carbon by microbes over direct oxidative degradation. This effectively reduces the energy consumption during wastewater treatment and enhances the recovery of nitrogen and phosphorus in the wastewater. When organic carbon is present in the form of acetate, approximately 48% of the organic carbon is degraded through aerobic microbe growth, while the remaining 52% of organic carbon is stored in the form of PHAs.

(2) In the targeted regulation model, the effluent meets the discharge requirements of the Class 1A standard of China. Given that the effluent water quality satisfies the relevant standard, carbon recovered in the form of CH_4 accounts for approximately 24.1% of the total carbon in the WWTP influent, while approximately 27.7% of the phosphorus in the WWTP influent can be recovered in the form of struvite.

(3) With regard to the total energy consumption and given that the effluent satisfies the water quality requirements of the Class 1A standard, the average energy consumption of the conventional model is 0.53 kWh/m^3, while the corresponding value for the targeted regulation model is 0.20 kWh/m^3, which is only 36.8% of the energy consumed in the conventional model. From the energy balance perspective, for every cubic meter of wastewater treated, an energy input of 0.31 kWh is required in the conventional model; however, a net energy recovery of 0.34 kWh can be achieved with the targeted regulation model.

(4) The total GHG emissions in the targeted regulation and conventional models are 0.312 and 0.545 kg CO_2-e/m^3, respectively. Therefore, compared with the conventional model, the GHG emissions can be reduced by 41.7% by adopting the targeted regulation model.

(5) Based on the process flow of the wastewater treatment, the total reactor volume for the conventional model is 1.86×10^5 m^3, while the total reactor volume for the targeted regulation model is 0.24×10^5 m^3, which indicates an 80% reduction in the total volume of the wastewater treatment reactors. However, with regard to the process flow of the sludge treatment, because the targeted regulation model involves the addition of a carbon transformation reactor, the total reactor volume of the sludge treatment process is 2.11×10^5 m^3, which is almost 8 times the total reactor volume in the conventional model. Consequently, the total volume of the main reactors and tanks of the targeted regulation model is 1.1 times that of the conventional model.

References

Ahn, Y. H., & Speece, R. E. (2006). Elutriated acid fermentation of municipal primary sludge. *Water Research, 40*(11), 2210–2220.

Doi, Y., Kunioka, M., Nakamura, Y., & Soga, K. (1987). Biosynthesis of copolyesters in Alcaligenes-eutrophus H16 from C-13-labeled acetate and propionate. *Macromolecules, 20*(12), 2988–2991.

Guisasola, A., Pijuan, M., Baeza, J. A., Carrera, J., Casas, C., & Lafuente, J. (2004). Aerobic phosphorus release linked to acetate uptake in bio-P sludge: Process modeling using oxygen uptake rate. *Biotechnology and Bioengineering, 85*(7), 722–733.

Jiang, Y., Chen, Y., & Zheng, X. (2009). Efficient polyhydroxyalkanoates production from a waste-activated sludge alkaline fermentation liquid by activated sludge submitted to the aerobic feeding and discharge process. *Environmental Science and Technology, 43*(20), 7734–7741.

Kim, H. W., Han, S. K., Song, Y. C., Baek, B. C., Yoo, K. S., Lee, J. J., et al. (2003). *Improved anaerobic biodegradation of biosolids by the addition of food waste as a co-substrate*. Trondheim: Norwegian University of Science Technology.

Lemos, P. C., Viana, C., Salgueiro, E. N., Ramos, A. M., Crespo, J., & Reis, M. A. M. (1998). Effect of carbon source on the formation of polyhydroxyalkanoates (PHA) by a phosphate-accumulating mixed culture. *Enzyme and Microbial Technology, 22*(8), 662–671.

Makinia, J. (2010) *Mathematical modelling and computer simulation of activated sludge systems*. IWA Publishing.

Metcalf, I., & Eddy, H. (2003). *Wastewater engineering: Treatment and reuse*. New York: McGraw-Hill.

Pijuan, M., Saunders, A. M., Guisasola, A., Baeza, J. A., Casas, C., & Blackall, L. L. (2004). Enhanced biological phosphorus removal in a sequencing batch reactor using propionate as the sole carbon source. *Biotechnology and Bioengineering, 85*(1), 56–67.

Qu, B. (2008). *Prodocution of biodegradable thermoplastics—PHA from excess sludge*. Beijing: Graduate University of the Chinese Academy of Sciences.

Sandor, B., Zsuzsanna, L., Horvath, Z. H., Gabor, S., & Cecilia, H. (2011). Comparison of the effects of microwave irradiation with different intensities on the biodegradability of sludge from the dairy- and meat-industry. *Bioresource Technology, 102*(2), 814–821.

Shizas, I., & Bagley, D. M. (2004). Experimental determination of energy content of unknown organics in municipal wastewater streams. *Journal of Energy Engineering-Asce, 130*(2), 45–53.

Talaro, K. P. (2008). *Foundations in microbiology: basic principles*. Columbus: McGraw-Hill Co.

Thomas, M., Wright, P., Blackall, L., Urbain, V., & Keller, J. (2003). Optimisation of Noosa BNR plant to improve performance and reduce operating costs. *Water Science and Technology, 47*(12), 141–148.

Tong, J. A., & Chen, Y. G. (2007). Enhanced biological phosphorus removal driven by short-chain fatty acids produced from waste activated sludge alkaline fermentation. *Environmental Science and Technology, 41*(20), 7126–7130.

Wang, D. B., Li, X. M., Yang, Q., Zheng, W., Liu, Z. Y., Liu, Y. L., et al. (2009). The probable metabolic relation between phosphate uptake and energy storages formations under single-stage oxic condition. *Bioresource Technology, 100*(17), 4005–4011.

Wang, X., Liu, J., Qu, B., Ren, N.-Q., Qu, J. (2013). Role of carbon substrates in facilitating energy reduction and resource recovery in a traditional activated sludge process: Investigation from a biokinetics modeling perspective. *Bioresource Technology, 140*, 312–318.

Wang, x., Daigger, G., Lee, D.-J., Liu, J., Ren, N.-Q., Qu, J.,Liu, G., Butler, D. (2018). Evolving wastewater infrastructure paradigm to enhance harmony with nature. *Science Advances, 4*(8), eaaq0210.

Wiesmann, U., Choi, I. S., & Dombrowski, E.-M. (2007). *Fundamentals of biological wastewater treatment*. Weinheim: Wiley-VCH Verlag GmbH & Co. KGaA.

Zeng, R. J., Yuan, Z., & Keller, J. (2006). Effects of solids concentration, pH and carbon addition on the production rate and composition of volatile fatty acids in prefermenters using primary sewage sludge. *Water Science and Technology, 53*(8), 263–269.

Zhang, P., Chen, Y., Zhou, Q., Zheng, X., Zhu, X., & Zhao, Y. (2010). Understanding short-chain fatty acids accumulation enhanced in waste activated sludge alkaline fermentation: Kinetics and microbiology. *Environmental Science and Technology, 44*(24), 9343–9348.

[illegible]

Chapter 8
Conclusions and Prospects for the Future

8.1 Conclusions

In the present work, the inadequacies of wastewater treatment processes were assessed using the life cycle assessment method, which has been widely applied in many fields. Based on our study, we propose the following: (1) resource recovery of organic matter, nitrogen, and phosphorus in municipal wastewater for the reduction of environmental impacts; (2) a novel assessment system based on the concepts of minimization of environmental impacts and maximization of resource recovery and utilization; and (3) a targeted regulation model of wastewater treatment aimed at energy conservation, reduced carbon emissions, and resource recovery based on the use of transfer and transformation of organic matter in wastewater as means of regulation. The following conclusions can be drawn:

(1) The chemical consumption, energy consumption, and greenhouse gas (GHG) emission inventory of different wastewater treatment processes considerably differ. From the perspective of life cycle environmental impacts, a stringent wastewater discharge standard increases the complexity of wastewater treatment processes and aggravates the combined environmental impacts generated by wastewater treatment, although it assures the quality of the water environments of the receiving waters. To address this issue, a novel assessment framework for the multi-objective management of wastewater treatment is proposed in the present work. The results of the assessment of identical wastewater treatment processes indicate that the resource recovery of organic matter, nitrogen, and phosphorus in wastewater effectively reduces the impacts of wastewater treatment on the environment. In addition, environmental impacts do not necessarily increase when the wastewater discharge standard becomes more stringent.
(2) The weights of the indicators reflect the key points of a particular period. The relative weights of various indicators for different countries (regions) and different historical periods significantly differ. In general, the weight of the same indi-

X. Wang, *Energy Consumption, Chemical Use and Carbon Footprints of Wastewater Treatment Alternatives*, Springer Theses,
https://doi.org/10.1007/978-981-13-5983-5_8

cator in developing and developed countries significantly differs and exhibits certain patterns over the years. For instance, the weight of bioenergy recovery in developed countries shows a significant increase between 2000 and 2009, while the weight of bioenergy recovery in developing countries between 2000 and 2009 is largely similar to the trend for developed countries between 1990 and 2000. In other words, the weights of certain indicators in developing countries follow the trends of the corresponding weights in developed countries.

(3) Based on the use of statistical principles and historical data, weight prediction models were constructed in the present work. Scenario simulations and predictions of weight-based multi-objective management of wastewater treatment indicate that concurrent multi-resource recovery and utilization during wastewater treatment have positive environmental effects in both developed and developing countries. The resource recovery of phosphorus in the form of struvite is the most effective measure for the reduction of environmental impacts of wastewater treatment in developed countries. The nutrient recovery via sludge composting is the most effective operation strategy for the achievement of sustained development of wastewater treatment in developing countries.

(4) The type of organic substrate in wastewater significantly influences the energy consumption levels and resource recovery potential of pollutants during the wastewater treatment process. It must be pointed out that the storage of organic matter by microbes is enhanced and the proportion of organic matter that is directly oxidatively degraded is reduced due to the use of acetate as substrate in the wastewater treatment system. This effectively reduces the energy consumption of wastewater treatment and promotes the recovery of phosphorus in wastewater.

(5) Compared with the conventional wastewater treatment model, the targeted regulation model attains 62.3% savings in the operation energy consumption, a 41.7% reduction in total GHG emissions, and zero consumption of exogenous carbon sources and phosphorus removing agents under the premise that the effluent satisfies the water quality requirements of the Class 1A standard of China. In addition to the maintenance of the energy self-sufficiency of operations, the target regulation model also achieves a net energy recovery of 0.34 kWh for every cubic meter of treated wastewater.

8.2 Recommendations and Outlook

The main novelties of the present study are the resource recovery of organic matter, nitrogen, and phosphorus in municipal wastewater for the reduction of the environmental impacts of wastewater treatment; the proposal of a novel assessment system based on the concepts of minimization of environmental impacts and maximization of resource recovery and utilization; and the subsequent construction of an optimized regulation model of wastewater treatment. However, this study is still in its initial development stage. Although the establishment of assessment methods and construction of regulation models have been explored to a certain extent, this study is far from being exhaustive. Therefore, to apply the study results, an in-depth investigation regarding the following aspects is required:

(1) Optimization of the assessment system for the multi-objective management of wastewater treatment including the continued development of the assessment framework for resource recovery of pollutants with respect to wastewater reuse and waste heat recovery; environmental impact analysis of gaseous pollutants in wastewater treatment; proposal of a corresponding assessment indicator system and assessment criteria; construction of an assessment system aimed at the control of gaseous pollutants; introduction of an ecological risk assessment system for resource recovery of pollutants; and establishment of methods for the assessment of potential ecological hazards.
(2) Experimental studies of the targeted regulation model for wastewater treatment should be carried out to determine the optimum operation parameters and control conditions of the regulation model; concurrent analysis of the biochemical kinetic processes of key biological units in the regulation model to investigate the influence of microstructures and evolutionary processes of microbes on these biological units and analyze the biological effects of targeted transfer and transformation of pollutants within these biological units; execution of research, development, and technological integration for novel reactors.

Structural regulation models have been explored to a certain extent, this study is far from being exhaustive. Therefore, to apply the study results, an in-depth investigation regarding the following aspects is necessary:

MIX
Papier aus verantwortungsvollen Quellen
Paper from responsible sources
FSC® C105338

If you have any concerns about our products, you can contact us on
ProductSafety@springernature.com

In case Publisher is established outside the EU, the EU authorized representative is:
Springer Nature Customer Service Center GmbH
Europaplatz 3, 69115 Heidelberg, Germany

Printed by Libri Plureos GmbH
in Hamburg, Germany